The story of MATHEMATICS

Alistair Ross

Adam & Charles Black · London

Black's Junior Reference Books
General Editor R J Unstead

1 Houses
2 Fish and the sea
3 Travel by road
4 The story of aircraft
6 Mining coal
7 The story of the theatre
8 Stars and space
9 Monasteries
10 Travel by sea
12 Musical instruments
13 Arms and armour
14 Farming in Britain
15 Law and order
16 Bicycles
17 Heraldry
18 Costume
19 The story of the post
21 Castles
23 Coins and tokens
24 Fairs and circuses
25 Victorian children
26 Food and cooking
27 Wartime children
28 When the West was wild
29 Wash and brush up
30 Getting about in towns
31 Canal people
32 The story of oil
33 Home sweet home
34 Outdoor games
35 Horses
36 Earthquakes and volcanoes
37 Going to school
38 The story of mathematics

Ross, Alistair
The story of mathematics.—(Black's junior reference books)
1. Mathematics—1961–
I. Title
510 QA39.2

ISBN 0–7136–2424–8

Published by A & C Black (Publishers) Limited
35 Bedford Row, London WC1R 4JH

First Published 1984

ISBN 0-7136-2424-8

Typeset by August Filmsetting, Haydock, St Helens.
Printed in Great Britain by R J Acford Ltd, Chichester, Sussex

Contents

The Description and Use
OF THE
CARPENTERS-RULE:
Together with the Use of the
LINE of NUMBERS
(*Inscribed thereon*)
In Arithmatick and Geometry.
And the Application thereof to the Measuring of *Superficies* and *Solids*, *Gaging* of Vessels, *Military Orders*, Interest and Annuities: with Tables of Reduction, *&c.*
To which is added,
The Use of a (portable) *Geometrical Sun-dial* with a *Nocturnal* on the backside, for the exact and ready finding the hour of the Day and Night: And other Mathematical conclusions.
Also of a UNIVERSAL DIAL for the use of *Seamen* or others.
With the use of *Sliding* or *Glasiers-rule.*
Collected and fitted to the meanest capacity
By *J. B.*
London, Printed by *W. G.* for *William Fisher* at the *Postern-gate* near *Tower-hill*, 1662.

1 Starting to count

One, two, three, four, five –
Once I caught a fish alive.
Six, seven, eight, nine, ten –
Then I threw him back again.

Why did you let him go?
Because he bit my finger so.
Which finger did he bite?
This little finger on the right.

Counting rhymes like this have been used for hundreds of years to help young children learn to count. Being able to count and work out problems with numbers and shapes helps us to describe and understand the world and people around us.

Once, long ago, people probably didn't need to count. Early hunters and food gatherers would have wanted to know if they had enough food, or if they could reach water and shelter. But they probably only needed a few counting words like 'one', 'two' and 'many'.

One of the first things we learn in school is how to count and to use numbers

Counting cattle in Thebes, 1400 BC. Scribes record the numbers reported by herdsmen

As life became more organised, people needed to count and to use numbers, so that they could build shelters, plan crops, and share out the harvests.

Many groups of people started to count using their fingers and toes. If the words used for some numbers are translated, they mean things like 'whole hand' (five), 'and one on the foot' (eleven), 'my hands and my feet' or 'man' (twenty), and 'two men' (forty).

This is why many counting systems are based on groups of five or ten. Counting in groups saves having lots of different words for new numbers.

This system, used by the Joloff in Africa, counts in groups of five.

1	wear	6	judom wear
2	yar	7	judom yar
3	yat	8	judom yat
4	yanet	9	judom yanet
5	judom	10	fook

It is easy to work out how the Joloff count up to nineteen. Counting in base five is a bit like using a tally chart.

Two different ways of counting hours. On the clock the hours are counted in base 12, but base 24 is used on the board

In English, we count mostly in base ten. But we still keep some special words for eleven to thirteen (not onety-one to onety-three), and for twenty and thirty (not twoty and threety). The counting system in Gujarati is also based on ten, but the new word for each group of ten comes just a bit earlier than in English.

20	wyss	26	chherwyss
21	ekwyss	27	styawyss
22	bauwyss	28	utyawyss
23	oltherwyss	29	ongtryss
24	chharwyss	30	tryss
25	pachwyss		

Many different bases have been used for counting. Base two, or binary arithmetic, is used in computers. We count and group seconds and minutes in sixties. Hours are sometimes counted in base twenty-four.

Here are two ways of counting used in Britain. The first is the way some Yorkshire shepherds used to count their sheep. The second is counting in Welsh. What base is being used?

yan tyan tethera methera pimp	sethera lethera havera dovera dick	yanandick tyanadick tetheradick metheradick bumfit	yananbumfit tyanabumfit tetherabumfit metherabumfit jiggit

un dau tri pedwar pump	chwech saith oeth naw deg	unddeg dauddeg triddeg pedwarddeg bumddeg	un ar bumddeg dau ar bumddeg tri ar bumddeg pedwar ar bumddeg ugain

The first ten numbers are in base ten, but then they change to base five.

People could use their fingers for counting up to quite high numbers. For example, in base five, it is possible to count up to five on the fingers of the right hand, and use each finger of the left hand to show a right-hand full (a group of five). If a second person helped, each finger of their right hand could count the left-hand-fulls (a group of twenty-five), and their left hand could count groups of 125. So two people could count up to 625 in this way.

When very large numbers were needed, or several numbers were needed at the same time, finger counting was clumsy. People developed ways of writing numbers.

This stone carving from ancient Mexico records an important astronomical event

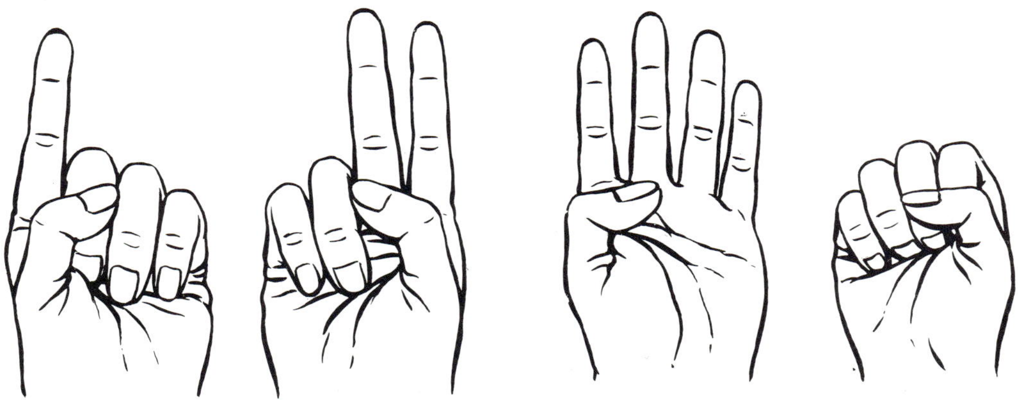

What number are these two people counting in base five?

Cuneiform writing

Written systems for numbers can be very different. The ancient Maya people of Central America based their numbers on their calendar. Twenty days (kins) made one uinal, and eighteen uinals made one tun, or year. Twenty tuns made a katun (7,200 days), and twenty katuns made a cycle (144,000 days). The numbers looked like a kind of tallying.

one	𝋡	ten	𝋪
two	𝋢	eleven	𝋫
three	𝋣	twelve	𝋬
four	𝋤	fifteen	𝋯
five	𝋥	sixteen	𝋰
six	𝋦	nineteen	𝋳
seven	𝋧		

The number of uinals would be written above the number of kins. So twenty-one (one uinal and one kin) would be written as one twenty and one unit.

(one 20) 𝋡
(one unit) 𝋡

To show that in numbers such as twenty there were no kins, a special symbol was used.

(one 20) 𝋡
(no units) 𝋠

The number 408 would be written as one tun (360), two uinals (2 × 20 = 40), and 8 kins.

𝋡
𝋢
𝋨

Can you work out how we would write these numbers in the system we use?

𝋦	𝋭	𝋤
𝋫	𝋰	𝋠
𝋯	𝋠	𝋲

The ancient Mexicans recorded astronomical events. The bars and dots are numbers

The Mayan system ended about a thousand years ago. The Chinese numbering system is one of the oldest we know of, and is still used today.

one	一	four	四	seven	七	ten	十
two	二	five	五	eight	八	hundred	百
three	三	six	六	nine	九	thousand	千

In this system, twelve is shown as ten *and* two, while twenty is two *lots of* ten. If the two is below the ten, it is added. If the two is above the ten, it is multiplied.

十二　　二十

How would you write these Chinese numbers in our system?

二千　　二千　　百九

三百　　二十

二十二

2 Measuring time

Numbers were first used for counting. Another important use for numbers was to measure time.

The first farmers probably planted and harvested their crops when the weather seemed right. If they planted spring wheat on the first fine day at the end of winter, they might be planting on an unusually warm winter's day and the crops would almost certainly be ruined by frosts. If the farmers waited until they were sure that winter was over, they might not plant in time to harvest before the autumn storms. Calendars were needed so that the farmers could work out more accurately when to plant and harvest.

Early people must have noticed the pattern of the seasons and tried to match it with the patterns of the sun, the moon and the stars.

Counting in days was probably the easiest way to measure time. If a farmer found the best time to sow his crops, and counted 365 days (about a year) until he next sowed, he would be sowing his crops at the same season as he had the year before.

Numbers are recorded on these two Sumerian stone tablets

The main line of stones points to where the sun rises on Midsummer's day

A solar year isn't exactly 365 days long. It takes the earth about $365\frac{1}{4}$ days to travel around the sun. If the farmer measured a year in 365 days, after four years he would be sowing a day too early. That wouldn't matter much, but after forty years his children would be sowing ten days early, and after a hundred years his grandchildren would be planting crops three and a half weeks too soon.

Some people based their calendars on the sun. By carefully noting where the sun rose each day throughout the year, the exact moment of the mid-year could be found. At midday on Midsummer's day, the sun reaches its highest point in the sky for the whole year. At midday in midwinter, the sun is at its lowest point in the sky. If the farmer always counted on from these days, he would be sowing at the same time each year. Some years would be 366 days long, and some would be 365 days long, but that wouldn't matter because the seasons would be right.

Stone circles, like Stonehenge, were probably built to mark the important days of the year. The main line of stones at Stonehenge points exactly to where the sun rises on Midsummer's day. Other lines of stones may point to sunrise and sunset on midwinter's day.

Calendars were needed so that farmers could work out when to plant and harvest. These farmers in ancient Egypt depended on the priests to interpret the calendar

Often, the secrets of the calendar were known only to the priests. In ancient Egypt, it was important for the farmers to know when the river Nile would flood. The priests had observed that the flood came on a particular day each year. This was the first day that Sirius appeared in the sky at sunrise. Astronomers were able to predict when this would happen, but the priests then performed ceremonies which they said caused the flood. Of course, they also charged the poor farmers a fee to perform these ceremonies.

The American Indians based their calendar on the moon. The moon changes its visible shape night by night. It changes from a new moon to a full moon in about 15 days and then back again in about another 15 days. So the American Indians measured time in months which each lasted 30 days. But a lunar month isn't exactly 30 days long. It lasts just over $29\frac{1}{4}$ days, which isn't an easy number to count with.

Thirty-day months do not fit exactly into a year. Twelve months based on a 30-day month would be $5\frac{1}{4}$ days less than a solar year, but thirteen months would be $24\frac{3}{4}$ days too long.

The Islamic calendar is also based on the moon. To solve the problem of having $29\frac{1}{4}$ day months, the Islamic calendar has twelve months of 29 days and 30 days alternately. This makes up a year of 354 days, which means that 33 Islamic years are roughly equal to 32 Western years. The fasting month of Ramadan is held during the ninth month of the year. So each year, Ramadan is eleven days earlier than the year before, when counted by the Western calendar.

The Babylonians and Greeks tried to solve the problem of matching the seasons by developing a calendar based on both the sun and the moon. Every month started with a full moon. Some months were 29 days long and some 30 days long. Seven out of every 19 years had to be given an extra month to catch up with the seasons.

The Romans used the Greek calendar at first. Some of the names of their twelve months are familiar, though the Romans began their year with Martius: Martius, Aprilus, Maius, Junius, Quintilis, Sextilis, September, October, November, December, Januarius, Februarius.

By this time, the calendar was in a bit of a mess. To make the calendar fit in with the seasons, an extra month was sometimes added after Februarius, at the end of the year. These years would be 377 or 378 days long. Unfortunately, the Roman leaders tended to add the extra months when it suited them, to make a politician's term of office last longer, or to change the date of an election. By 190 BC, after about 500 years, the calendar was 117 days out!

Julius Caesar called for expert advice from the Greek astronomers. They gave the year 46 BC an extra two months, to catch up with the sun. This 445 day year was called the Year of Confusion. Then the year was made ten days longer by adding an extra day to ten of the twelve months. Every fourth year, an extra day was to be added to Februarius. The year was to start in Januarius, not Martius, and the month Quintilis was renamed Julius in Caesar's honour. The Julian calendar was very similar to the Western calendar of today, with 365 days in the year, except for leap years which have 366 days.

Roman calendar with months, dates and days of the week. The three pegs can be moved from hole to hole

After Julius Caesar's death, the Romans still managed to make a few mistakes. The leap years were put in every third year, instead of every fourth year. Finally, in AD 8, Emperor Augustus had to sort things out, and the month named Sextilis was changed to Augustus in his honour.

Although the new system worked well for many years, the solar year is not exactly $365\frac{1}{4}$ days long. It is one minute and 14 seconds shorter. By AD 710, the English monk, Bede, noted that the seasons were changing about three days earlier than they had 400 years before. Again, people began adding extra days to the year. But this time, with no Emperor to insist that everyone made the same changes, different countries, and even different cities, were using different systems.

Hogarth's engraving of *The Election* shows a poster of 'Give us back our 11 days'

Pope Gregory XIII

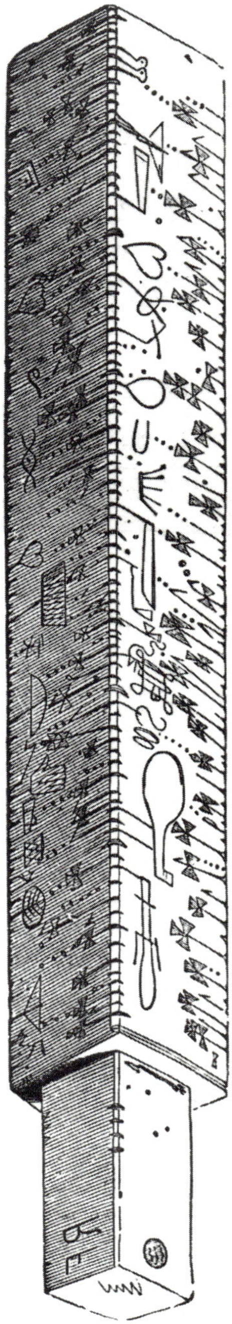

Saxon calendar

In 1582, Pope Gregory XIII decided to make new rules that would catch up ten days on the Augustine calendar. In that year, the day after the 5th October was to be the 15th October. It was also ruled that leap years would not occur in the century years (1700, 1800, and so on) unless the century year could be divided exactly by 400 (such as 1600 or 2000).

Unfortunately, the Gregorian calendar was only accepted in Roman Catholic countries. Protestant countries, like England at the time, refused to accept the change at first. Britain eventually changed in 1751, by which time the calendar was 11 days out. When the calendar was adjusted, many people felt cheated out of these eleven days, and there were even riots with the slogan, 'Give us back our eleven days!'.

We still use the Gregorian calendar today. The first 400 year cycle finished on 15th October, 1982. It should work for another 3 000 years before being even one day out.

Measuring small units of time was not usually very important for farmers. They could tell whether it was morning, noon or afternoon by looking at the sun. When people began to live in towns, they needed to know when markets would be open, and when meetings would be held. Usually, the simplest way to do this was to measure the position of the sun, using a shadow clock or a sundial. A sundial has an upright pointer, called the gnomon, which casts a shadow. The direction of this shadow shows the time. Sundials and shadow clocks were only accurate to within half an hour and were of no use at all on cloudy days.

Greek and Roman engineers developed water clocks which measured the steady drip of water from one container to another. The level of water in the bottom container showed how much time had passed since the clock was started. If you try this you will find that, as the water in the top container goes down, the drips get slower. To get round this problem the Roman clypsedra had a special reservoir to keep the level of water in the top container constant.

The ancient Chinese measured time by burning a candle on its side. As the candle burned at a constant rate, they could tell the time from the position of the flame. Sometimes, a string with weights tied to each end was laid over the candle. When the flame reached the string, the weights would fall and hit a gong. This was an early type of alarm clock.

Mosaic of Anaximander who is said to have invented the sundial

Our system of measuring time in seconds and minutes, and counting them in sixties, comes from the Babylonians. They discovered that if you divided a circle into six, this would give you six equal-sided, equal-angled triangles. Because the Babylonians recorded numbers in base 60, they divided each angle of the triangle into 60 degrees. So the angles at the centre of the circle totalled 360 degrees. The faces of early mechanical clocks were divided in the same way as the circle.

Mechanical clocks were not in common use in Europe until about the 13th or 14th century, and were not very accurate. It wasn't until the pendulum clock was developed that seconds and minutes could be measured accurately. In 1583, Galileo Galilei was watching a lamp swing back and forth in Pisa Cathedral and idly timing the swings against his own pulse. He noticed that the swings were regular, taking exactly the same length of time to swing from side to side. Galileo worked out that the time which a pendulum takes to swing from side to side depends, not on its weight or how far it swings, but on the length of the pendulum. The regular motion of the pendulum would be an accurate way of marking regular moments of time. Galileo didn't begin to apply his idea to clocks until he was an old man and he died before he had completed his work. It was another three-quarters of a century before the Dutch mathematician, Christian Huygens, built the first pendulum clock.

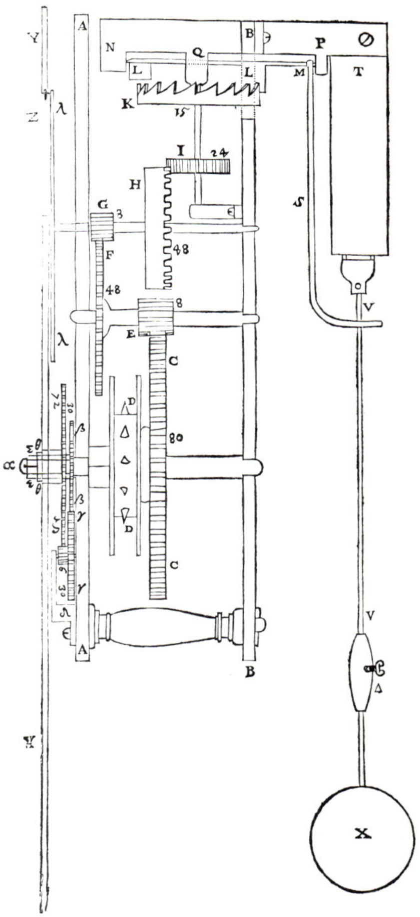

Diagram of Christian Huygens' first pendulum clock

A primitive shadow clock from Egypt

3 Shapes and patterns

An Egyptian scribe records measurements of land for taxes

The ancient Egyptians had a very important reason for being able to measure accurately. Every year, the Nile flooded the farmland along its banks. When the floods had gone down, the farmers had to redivide the land between them. The farmers had to agree upon the boundaries of their fields and also had to pay a tax on the amount of land they had enclosed.

Small units of length were based on the body. One cubit was the distance between the elbow and the tip of the middle finger. A palm, the width of four fingers, was one sixth of a cubit. A digit was a quarter of a palm. These measurements varied from person to person, so they were standardised using rules and lengths of string. These standard measures were so accurate that they could be used to build the Pyramids.

The Egyptian farmers marked the shape of their fields by measuring the angles. Right angles were measured by using a marker peg and a length of string like a compass, or by using a tightly pulled string to mark off a triangle with sides in ratios of lengths such as 3:4:5.

Larger measures of length were often based on farming. The mile was based on the Roman *mille passum* and was a thousand paces of five feet. A furlong meant 'a furrow long' and was based on the length of a furrow that a team of oxen could plough before stopping to rest. The mile was made into a standard eight furlongs by Elizabeth I, so that miles and furlongs could be used together.

As the first towns were built, people needed to plan more carefully how and where they put up buildings. For example, in a village, one small well might be enough to supply everyone with water. In a town, more than one well might be needed so that everyone could live close enough to a water supply.

To supply a town with water, canals, qanats (underground tunnels) or aqueducts often had to be built. This meant that the height of the land around the town, and the distance between the town and the water supply had to be measured. Careful calculations had to be made to build the canals, qanats and aqueducts. As long as 3500 years ago, engineers in ancient Persia constructed hundreds of underground qanats. The qanats could transport water from springs to villages up to 160 kilometres away. Some of these qanats are still in use today.

A mural at Abd el Qurna shows how land was measured in Egypt, 1400 BC

The Colosseum, Rome

As well as careful measurement, ideas of pattern and shape were important for building and planning. The ancient Greeks were interested in regular shapes, like the tetrahedron, which has equal sides and angles. Greek architecture was based on straight lines, right angles and equal proportions, and Greek buildings were often constructed from crossbeams supported on pillars.

Roman architecture was quite different. The ancient Romans invented the arch, where stones lock together to share their weight evenly between widely spaced pillars. Roman engineers calculated how to build aqueducts that could take water hundreds of miles. Complicated mathematics was needed to plan and build arched bridges and domed buildings.

The Romans were fascinated with shapes that fitted together, and many of their buildings had mosaic floors. The Romans called these mosaics *tessellae*, from which we get the word *tessellation* – fitting shapes together to form a regular pattern.

Simple regular shapes, like squares or hexagons, tessellate easily. Other tessellation patterns may need two shapes, such as octagons and squares.

Hindu and Arab artists created more elaborate patterns. Hindu Rangoli patterns are based on simple lines that are reflected or mirrored from different angles. Then the same pattern is repeated over and over again to form new shapes. Hindu and Sikh families use patterns like this to decorate their homes for important occasions, such as the yearly festival of Diwali.

Islamic patterns are often based on circles with geometric shapes drawn over them, such as triangles, trapezia or rhomboids. You can work out how these two Islamic patterns were constructed from the same basic shapes – circles and squares.

The Parthenon, Athens

Rangoli patterns

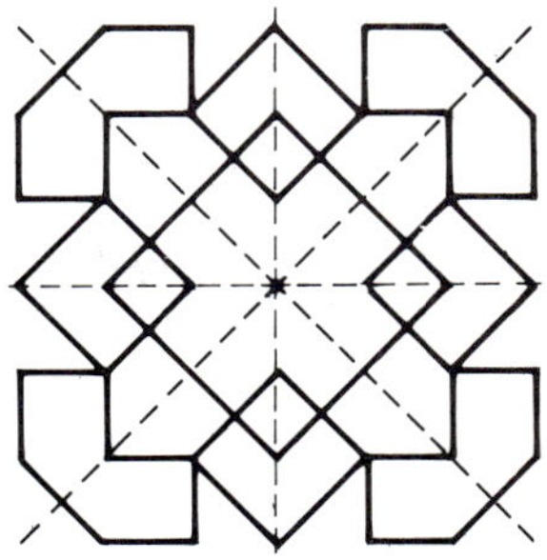

Islamic patterns

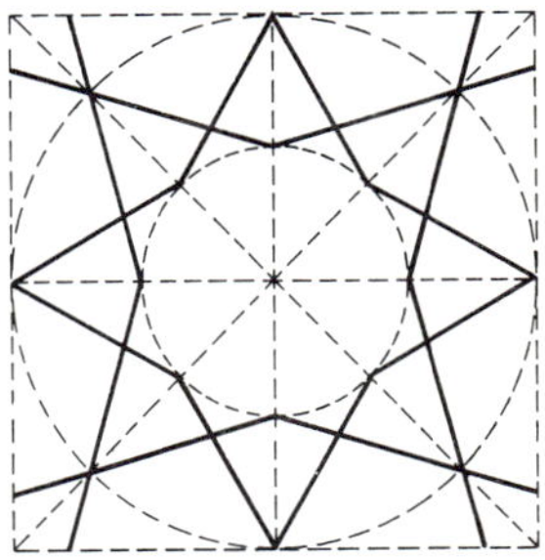

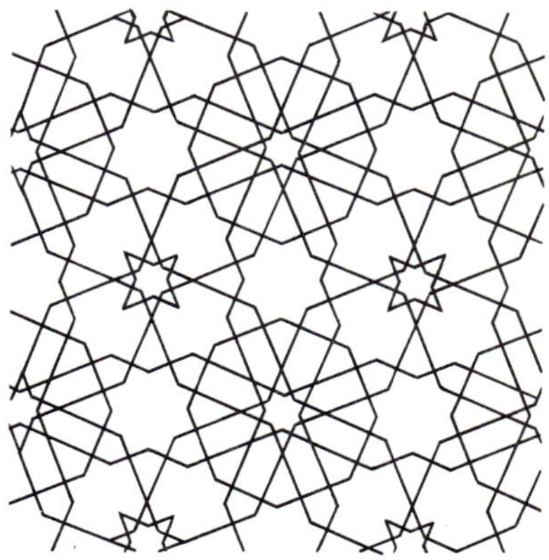

In Europe in the 14th to 16th centuries, artists began to develop new ideas about shape and size. Until this time, paintings were not very lifelike. Pictures usually showed important people or religious events and weren't supposed to be very realistic.

Artists then began to understand perspective and used the idea to make their pictures reflect what they actually saw more accurately. They realised that the further away an object is, the smaller it looks. They also realised that objects in the distance look as though they are closer together than they really are – they converge. For example, in Duccio's *The Last Supper*, the ceiling beams are really parallel. But by painting them at an angle, Duccio has given the room depth, and made it more realistic.

***The Last Supper*, Duccio di Buoninsegna**

Sketch for *The Adoration of the Magi*, Leonardo da Vinci (above)

***Perspective study of a Chalice*, Uccello (right)**

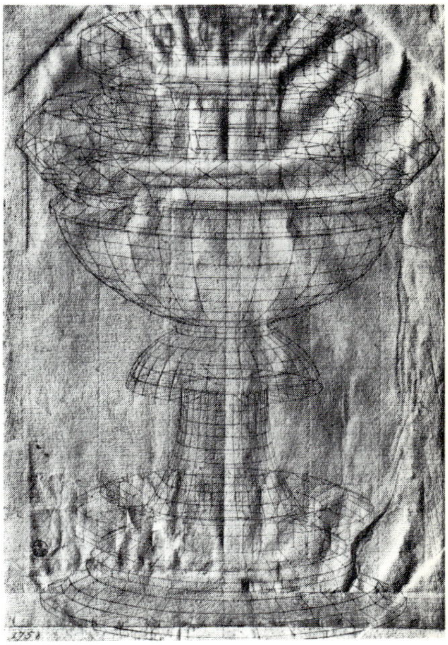

This idea can be seen even more clearly in Leonardo da Vinci's *Adoration of the Magi.* The building's lines all meet at one point, called the vanishing point, where everything in the distance eventually converges. Leonardo da Vinci was a mathematician, scientist and engineer, as well as an artist. Artists often worked out the mathematical details of their paintings before they began. For his *Perspective study of a Chalice,* Uccello made a sketch showing exactly how every curve and bend of the cup is plotted. It is said that Uccello worked long into the night on his sketch. When his wife finally called him to bed he remarked, 'How sweet a thing is this perspective.'

In 1528, a German artist named Albrecht Dürer wrote a book that explained the theory of perspective. He imagined the artist's canvas to be a glass screen, through which he saw the scene to be painted. By marking a grid on the screen, each of the objects in the scene to be painted could be plotted as the eye actually saw them. This is called the projection of a scene, and later became very important in the mathematics of map drawing.

Designer of the sitting man, **Dürer**

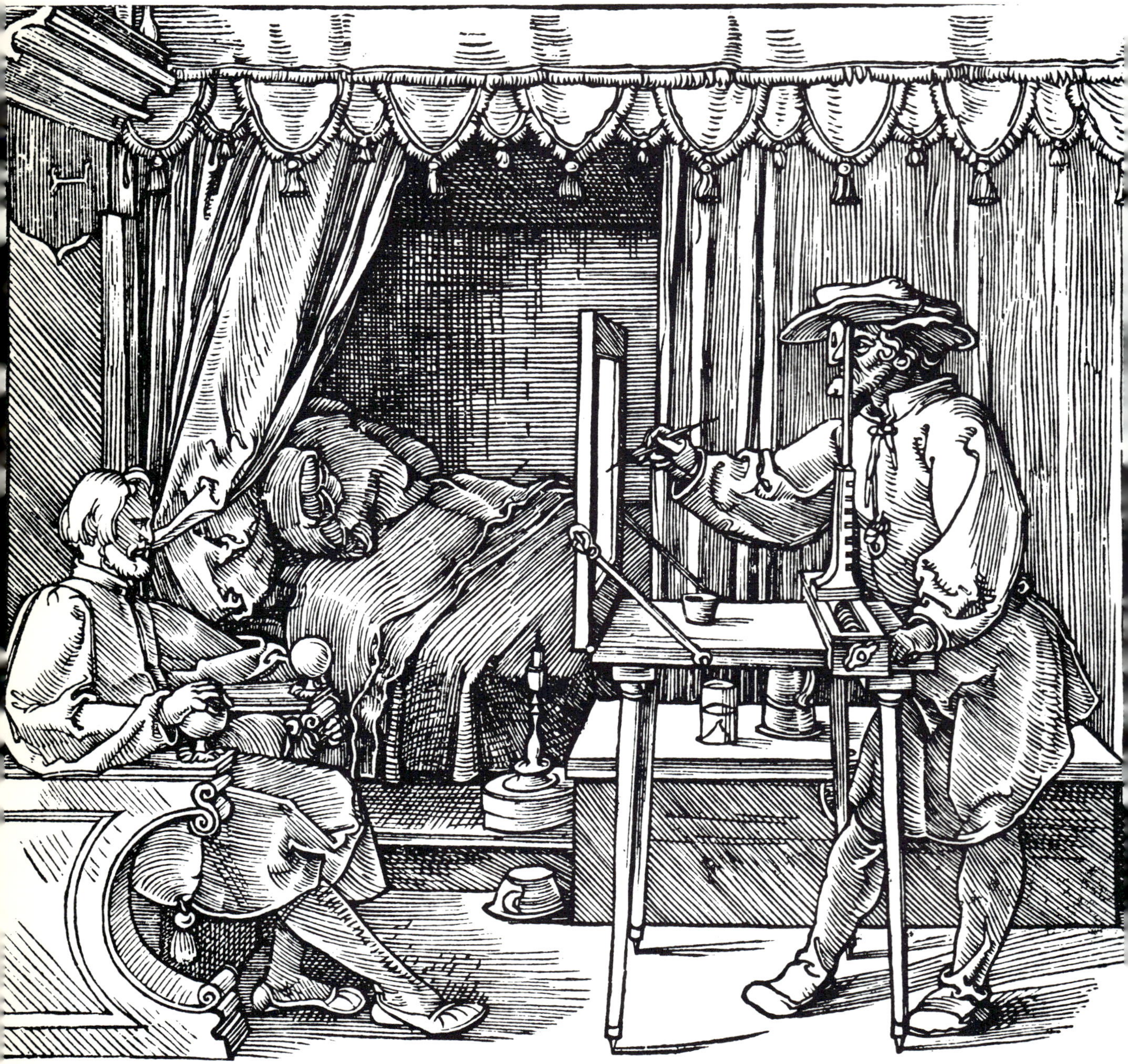

False Perspective, William Hogarth

Relativateit, Maurits Cornelis Escher

Of course, once discovered, the rules of perspective could be used to trick people. Hogarth's engraving, *False Perspective*, looks quite ordinary at first glance, but it contains lots of oddities. Written at the bottom of this engraving is, 'Whoever makes a design without the knowledge of perspective, will be liable to such absurdities as are shown in this print.' More recently the artist Escher has played an elaborate perspective trick in *Relativateit*. Try to find the bottom of the stairs. When is the picture the right way up?

4 Finding the way

As trading between countries increased, people began to travel more. Making short journeys by land wasn't usually a problem. A traveller could ask the way to the next town or village, and then ask again. Simple strip maps showed the route from one place to another and gave the various landmarks to look out for. But long distance travel could be difficult in areas where there were poor tracks and few people – and dangerous where there were wars or bandits.

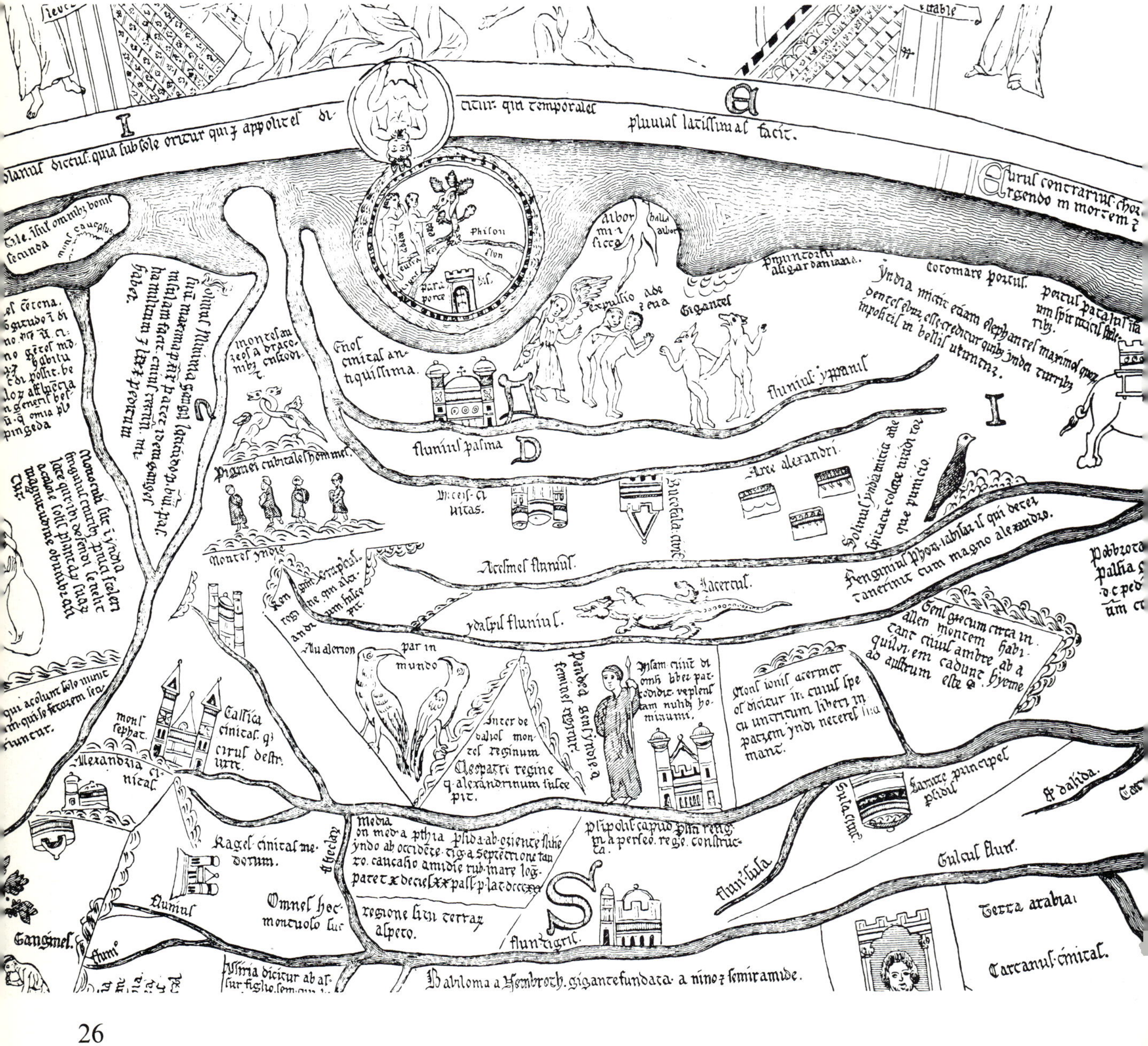

Hereford *Mappa Mundi* of Richard of Haldingham, 13th century

A Picture from a 16th century version of *The Travels of Marco Polo*

Travelling by sea could be dangerous, too. Sailors usually found their way by sailing around the coastline, always staying in sight of land. This could mean a very long journey and the sailors also ran the risk of running aground or of being seen from the shore by pirates. It would have been faster and safer to cut across the open seas in a direct line. Travellers needed to be able to calculate distances and directions.

Direction could be measured with a compass. The compass had been developed using a magnetized ore called lodestone which, when suspended on a string, always pointed north-south. Using a compass, direction could be measured either in degrees (clockwise from north), or by points of the compass (north-east, south-west, and so on). These measurements were used to calculate bearings. Early navigational maps, called *portolano*, were covered with networks of lines that gave bearings to follow. The navigator used a compass to head the ship along the map line that best fitted the route he wanted. He would follow this line until he reached land.

Early mariner's compass

Picture from a 16th century manuscript showing Ptolemy measuring the altitude of the moon with a quadrant. Ptolemy lived 1400 years earlier

This ship's sandglass will run for four hours

Ships were often blown off course and lost their bearings. The navigator needed to be able to measure what the ship's new position was. Using a sextant, the navigator could measure the position of the midday sun. The angle between the sun and the horizon could be used to calculate the ship's latitude. This was the ship's position north or south. Of course, the navigator had to know exactly when it was midday. Before accurate watches or chronometers were invented, sand glasses had to be turned over and over again during a long voyage. These were unlikely to be very accurate, even when three sand glasses were used and the best two readings taken.

Another way of calculating the latitude was to use an astrolabe to measure the position of the North Star. The latitude was found by measuring the angle between the North Star and the horizon.

Skilled mariners 'sailed latitudes'. They went north or south until they reached the line of latitude they wanted. Then they would sail east or west until they reached land. The directions for the West Indies from Britain were 'south till the butter melts, then due west'.

An Arabic astrolabe made in AD 1067

The Royal Observatory, Greenwich in the 17th century

Calculating the ship's position east-west, along lines of longitude, was more difficult. About 300 years ago, the English mathematician and scientist, Isaac Newton, calculated a theory about the movement of the planets, including the earth. This theory led to a way of calculating longitude.

Since the earth makes one complete clockwise turn every 24 hours, different places will see the sun at its highest point at different times. Midday (when the sun is at its highest point) will be four minutes later for each degree west that is travelled. Greenwich lies on the longitude zero degrees (0°). So, midday an hour later than at Greenwich will be somewhere on a line 15 degrees west of Greenwich (perhaps the west coast of Africa or Iceland). Three hours earlier will be 45 degrees east of Greenwich (perhaps Iran or Madagascar).

Sand glasses were far too inaccurate for this kind of measurement, and ordinary pendulum clocks were upset by the pitching and rolling of the sea. In 1714, the British government, keen to give its merchants an advantage over foreign merchants, offered a £20,000 prize for the design of an accurate clock or chronometer. The prize was won in 1764 by John Harrison whose chronometer was just five seconds out after it had been tested on a journey to Jamaica. Together, the chronometer, sextant and compass made navigation far more accurate.

Navigation was also helped by improved map-making. Showing an almost spherical shape, like the earth, on a flat chart is impossible to do exactly, though there are several different ways of attempting it. One system popular in the 16th century, was devised by Gerhardus Mercator. He used a mathematical formula which gave accurate directions, but not accurate distances. Mercator's projection made places in the far north and south look much larger than they really are, but showed accurate bearings as straight lines.

Babylonian map of the world, 700 BC

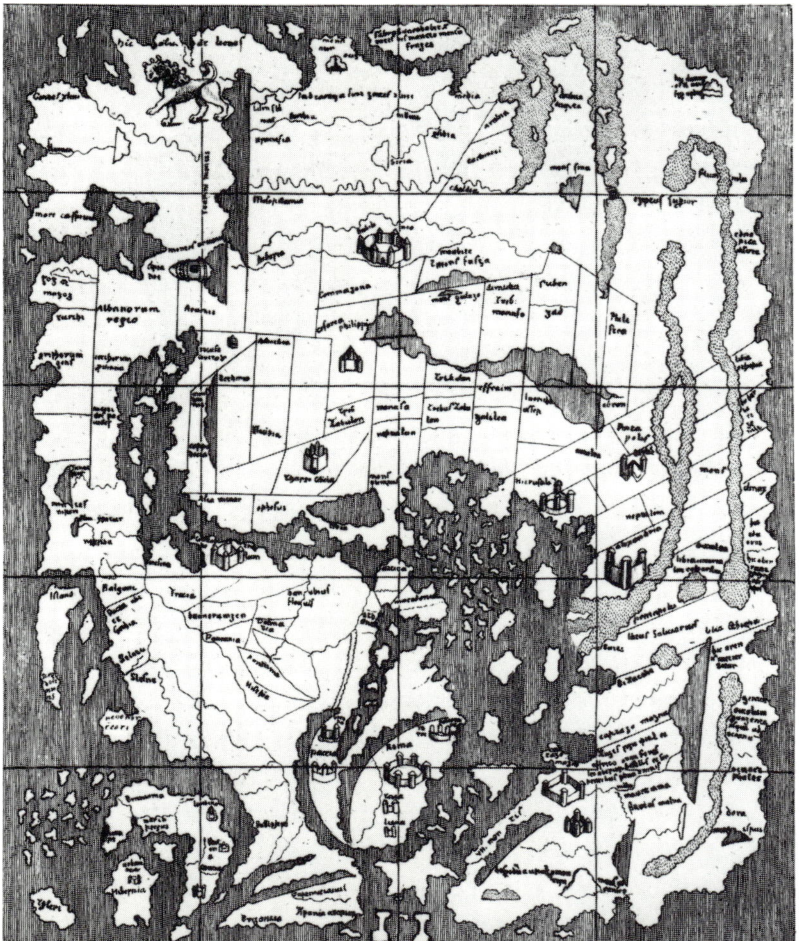

By the 10th century AD the map of the world was more detailed

Map of the world according to Ptolemy

Long before the telescope was invented, astronomers were trying to measure the movement of planets and stars. The Babylonians kept detailed records of the eclipses of the sun and moon. For eight centuries from 747 BC, the Babylonian astronomers used water clocks to time the eclipses. A cuneiform tablet (look at the picture on page 8) records an eclipse of the moon:

> 'Month 8, Day 14, at 12 minutes before sunrise – an eclipse began on the east side of the moon.'

That was 3rd November, 240 BC, and it has been calculated that an eclipse would have been seen in Babylon at that time.

The first total eclipse of the sun for which we have historical records, was recorded by Chinese astronomers on 17th July, 709 BC. In 221 BC, the Emperor Tsin-Chi-Hong-Ti ordered all scientific records to be destroyed. So many early Chinese records have been lost.

The Greek mathematician, Ptolemy, working in Egypt in about AD 150, used many of these observations to work out a system which explained how the planets moved. He believed that the sun, moon, planets and stars all moved in circles around the earth, each held up by its own crystal sphere. This system was accepted for hundreds of years.

People also believed that the stars had an influence on their lives and on their health. Doctors had to learn astrology – the art of predicting from the stars. Careful observations were made to find out just when the various stars would move across the skies. Then complicated patterns of star galaxies were plotted to try to predict the future.

The Islamic mathematicians of the Middle Ages were great astrologers, and made many of the observations used in the western world. They used astrolabes (look at the pictures on pages 29 and 63) to calculate when and where stars would appear in the skies. Wealthy Moslems often had astronomical calculators. These were tiny brass mechanical devices that could be turned to tell just when stars were due to appear.

In the 18th century, Jaisingh, ruler of Jaipur in India, built huge brick and cement instruments for his astronomical observatory. The reason for their enormous size was that he could then make much more precise observations than was possible with smaller brass instruments.

The observatory at Jaipur

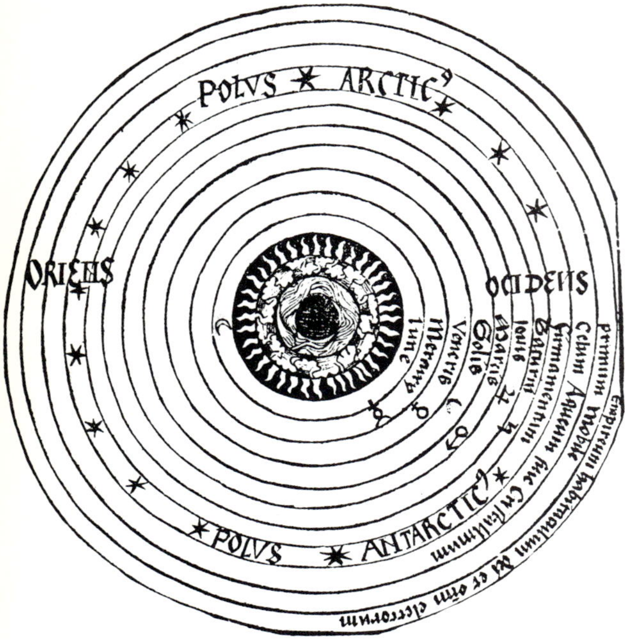

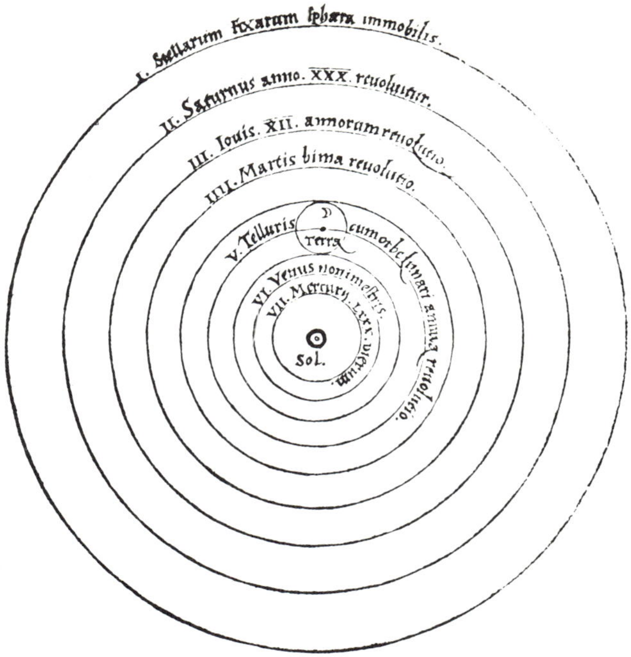

In Ptolemy's universe (left), the earth is at the centre of the heavenly spheres. By 1543, Copernicus had placed the sun at the centre (right).

The Ptolemaic system lasted for 1400 years and was used by both Christian and Islamic scientists. But Copernicus, Galileo and Newton had new ideas, made fresh observations and used new calculations. They showed that the planets, including the earth, moved around the sun and that the sun itself moved through space as one of a group of stars.

Copernicus was the first to upset the old system. He reasoned that the earth might as well revolve around the sun as the sun around the earth. Galileo used the newly invented telescope to observe the paths of the planet Venus and the moons of Jupiter. His observations supported Copernicus's theory. The way Galileo described his observations led him into trouble with the Roman Catholic church and he was made to take back what he had written. But his ideas spread across Europe, and led to new mathematical theories about how the planets and stars moved.

This work is still going on today, and fresh observations are being made all the time. Astronomers using powerful telescopes store new information about space in computers. The computers are able to use this information to make predictions which help in the study of atomic physics.

5 Trading and money

While artists and astrologers were busy developing complex mathematics to describe the world, more and more simple mathematics was being used in everyday life. As people became less self-sufficient, they began to bargain and trade with each other to fill their needs.

Trading goods, like wheat or oil, meant that quantities had to be agreed. These goods were often measured in pots or jars. Different countries developed different measures, and some merchants even used different measures for buying and for selling. The Greeks used a measure of about 39 litres which they divided into twelve amphorae. *Amphora* is the name of the measuring jar used for trading and is still used in Greece today. The Romans also measured in amphorae, but their measure was much larger than the Greeks' – about 36 litres.

In medieval Britain there were different gallons for measuring ale, corn and wine. The ale gallon was over a fifth bigger than the wine gallon. Gradually, inn-keepers made their ale measures smaller, and in 1824, Parliament had to lay down standard sizes.

Peasants at market, 1512, from a woodcut by Dürer

A 16th century market scene by Hans Bol

The Great Balance at Tournai Cathedral in the 15th century

The need to measure weights seems to have been even older than the need to measure quantities. Egyptian balances have been found that are over 7 000 years old. Weight systems varied from country to country, although the names were sometimes similar. The Roman pound was the same weight as one eightieth of a cubic foot of water. British weights, such as the hundredweight and the ton, all came from a rule made by King Alfred in about AD 850. He said that a 'weight' should equal a container of water that measured a hand-span in each direction. A hundred 'weights' became a hundredweight, and two thousand 'weights' became a ton.

All these weights and measures have been changed, and mostly replaced, by the metric system. This was first devised in 1795 after the French Revolution. The new French leaders wanted to replace the old system of weights and measures with a system that would be accurate enough for the new advances in science. They wanted the system to be based on standards that wouldn't change.

The French Academy of Science proposed a new unit of length based on the distance in a straight line from the North Pole to the Equator. One ten millionth part of this distance is a metre. One hundreth of a metre is a centimetre, and one thousand metres make a kilometre. The weight of a cube of water, one centimetre by one centimetre, is a gram, and a thousand grams make a kilogram. The volume of a kilogram of water is one litre. All the weights and measures are related to each other and all the units are based on the decimal system.

This Egyptian weighing scene, from about 2500 BC, is the earliest record of the use of balances for weighing

In this engraving made by Hogarth in 1740, the milkmaid holds out a tally stick and demands payment of the bill

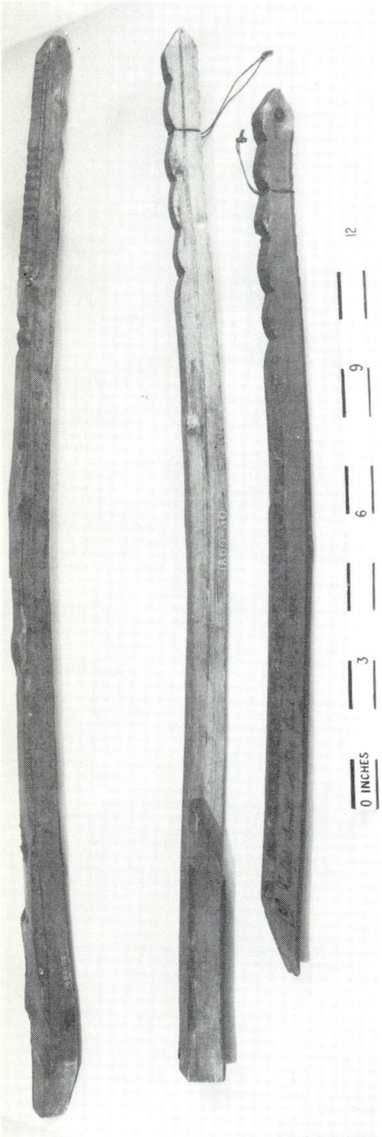

Tally sticks from the exchequer

At first, trade was based on bartering. An ox could be exchanged for so many measures of wheat, for example. But bartering didn't work very well if the person selling the ox wanted barley instead of wheat, or if they needed the wheat in six months' time.

For this kind of trade, people needed something that could buy all kinds of goods and services. Anything that had an agreed value would do. Some people used ox hides, and others used a particular kind of shell. Then, metal tokens shaped like ox hides were used instead. Each token stood for the value of one real ox hide and, of course, was much easier to handle. In this way, various coins were developed.

As trade increased, ways of recording transactions were needed. In England, wooden tallies were used as receipts and proofs of payment. Small nicks were cut in one side of a piece of wood to show the number of pennies, larger nicks showed the number of shillings, and a notch 'the size of a swelling barley corn' was cut for each pound. Cuts on the other side of the stick showed groups of pounds. The tally was then split in two. The buyer kept one piece (the stock) as proof of payment, and the seller kept the other (the wafer) as record. This system was used to record tax payments in Britain until 1812. The wafers for hundreds of years of taxes were kept in the basement of the Houses of Parliament, until in 1834, the wafers caught fire and burned the entire building down!

Coins made trading much easier for people using the same currency. When money was exchanged between countries with different currencies, more complex calculations were needed. Suppose that in one country a coin called a lira was worth one measure of wheat, or ten lira was worth an ox. In a second country, oxen might have been scarce, so the standard coin there, a franc, was worth one measure of wheat but twenty francs were worth an ox. When goods were traded between the two countries, how much were their coins worth? If they went by the value of wheat, one lira was equal to one franc, but if oxen were used for comparison, one lira was worth two francs. Merchants had to find ways of calculating the rate of exchange between different currencies.

These coins were minted in Israel in about AD 132

English merchant counting his money in the 15th century

Some calculations could be done on the fingers. One way of multiplying may go back as far as Roman times, and works for any pair of numbers between 5 and 10. For example, in the sum 7×9, take 5 from each number to be multiplied. Raise two fingers on one hand ($7-5=2$), and four fingers on the other hand ($9-5=4$). Add the number of raised fingers to give the number of tens ($4+2=6$, 60). Multiply the number of unraised fingers to give the number of units ($3 \times 1 = 3$). Add the tens and units together to get the answer 63.

For more complicated sums, a counting board or abacus would be used. The abacus seems to have been developed separately in Egypt, Babylonia, China, India, Japan and Peru.

Pythagoras sits on the right using a counting board. On the left, Boethius challenges with the new ways of calculating.

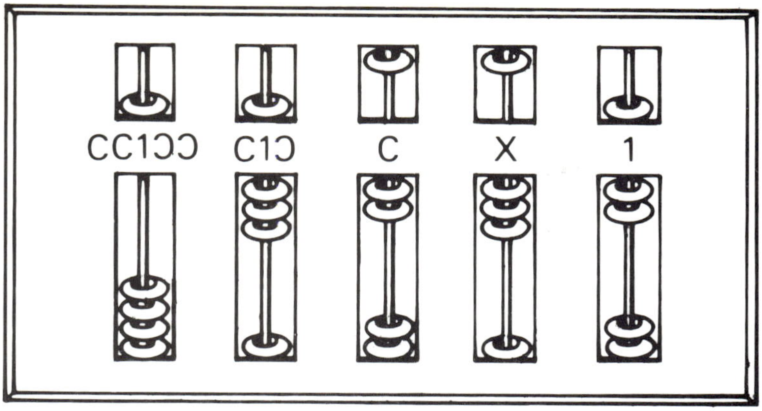

This Roman abacus shows 3 782. 2 is shown by the two counters in the bottom groove on the right that have been moved to the middle. The counter in the top groove hasn't moved so there aren't any fives. Can you work out the rest?

The Roman abacus had counters which were fixed in grooves. Each groove, from right to left, represented units, tens, hundreds, and so on. For every unit that was counted, a counter would be moved up on the bottom groove. For every five units counted, a counter would be moved up on the top groove.

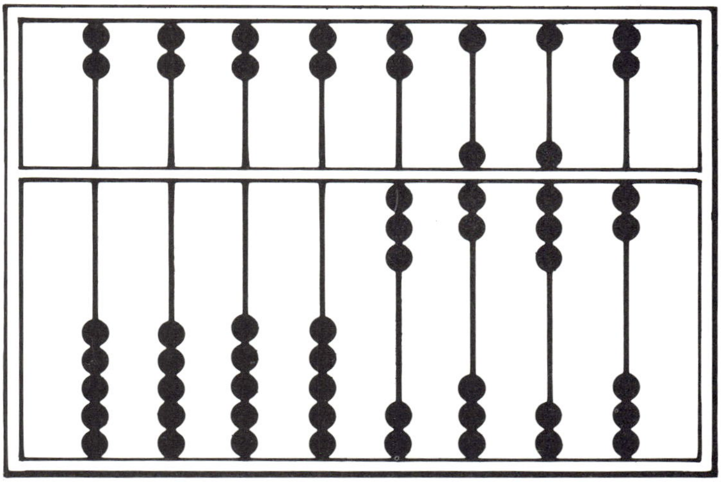

This Chinese abacus shows 3 782

The Chinese abacus is still used today, although in most far eastern countries the abacus has recently been replaced by the electronic calculator. The Chinese abacus is similar to the Roman abacus, except that beads are moved along bamboo rods, and are moved to the middle of the abacus instead of the top.

The Chinese word for calculation, *suan*, shows an abacus held from below by the symbol for a hand, and from above by the symbol for bamboo.

suan

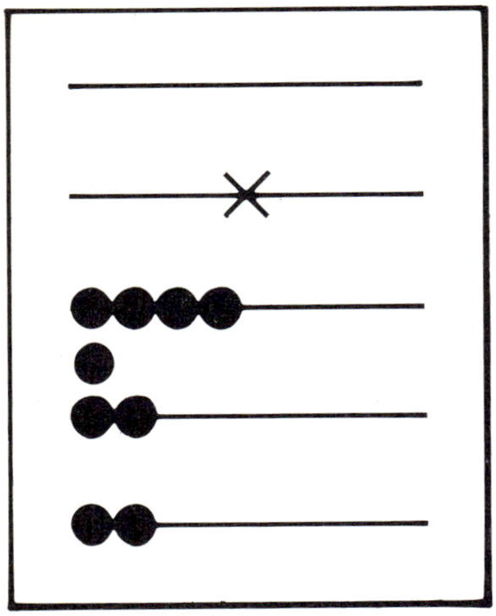

These counters show 472

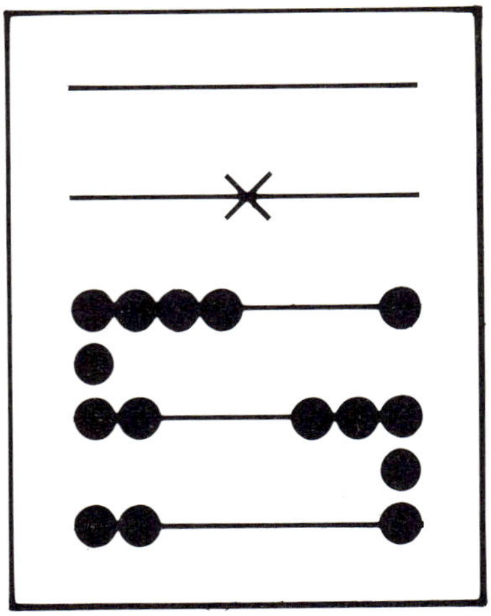

136 is added like this

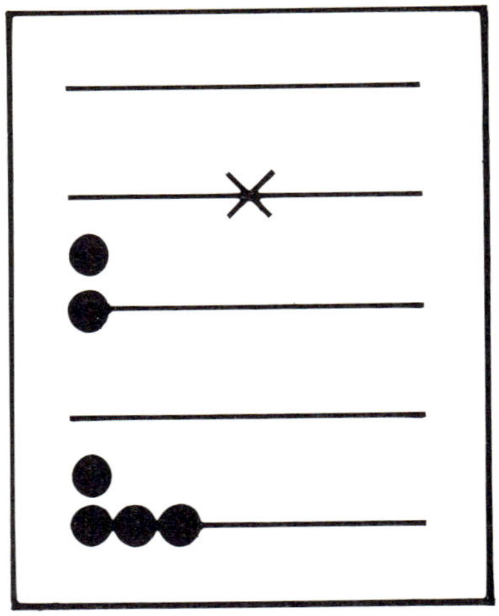

The answer is 608

Counting boards were common all over Europe in the Middle Ages. On the counting board there were lines for units, tens, hundreds and so on, on which counters were placed. Counters put in the spaces between the lines showed fives, fifties and so on.

The X on the thousands line was to help people remember which line was which. We still do this, but now we use a comma or a space instead of an X to separate thousands from hundreds.

The abacus and counting board could also be used to multiply and divide. At first, people multiplied numbers by adding the same number again and again. So 20×3 was found by adding $20+20+20$. Dividing numbers was done by a series of subtractions, so that $133 \div 33$ would be found like this:

$$
\begin{aligned}
133-33 &= 100 \quad (1)\\
100-33 &= 67 \quad (2)\\
167-33 &= 34 \quad (3)\\
34-33 &= 1 \quad (4)
\end{aligned}
$$

Therefore $133 \div 33 = 4$ and 1 remaining.

The Egyptian mathematician, Ahmes, worked out a quicker way of multiplying by doubling the figures. He would set about finding the answer to 20×6 like this:

$$
\begin{aligned}
1 \times 20 &= 20 \quad (1)\\
2 \times 20 &= 40 \quad (2)\\
4 \times 20 &= 80 \quad (3)
\end{aligned}
$$

Therefore $6 \times 20 = 2 \times 20 + 4 \times 20 = 40 + 80 = 120$.

The Italian mathematician, Gerbert (who became Pope Sylvester in AD 999), worked out some easier ways to divide using the abacus. Most mathematical calculations were carried out on the abacus or counting board and not written down on paper. Even the word 'calculate' comes from 'calculus' the word for a pebble or bead on an abacus. The counting boards used in shops became the shop counters of today.

All these developments helped merchants in medieval Europe and the Middle East to exchange goods in different currencies and trade was able to flourish between countries.

Keeping accounts at Rouen, 1466

6 Ways of calculating

In the past, writing down sums wasn't usually necessary. Using an abacus or counting board, most calculations could be done as they were needed. Then the answer could be written down.

Until about AD 1600, Europeans used Roman numerals to write down numbers. If you look at these numbers, you can see that working out a written sum would be very difficult. For example, we would write the sum 19 subtracted from 62 like this:

$$\begin{array}{r} 62 \\ -19 \\ \hline 43 \end{array}$$

Now try the same sum with Roman numerals and see what happens!

$$\begin{array}{r} \text{LXIII} \\ -\text{XIX} \\ \hline \end{array}$$

Calculations were almost impossible using Roman numerals. With our system of numbers, we can tell that the '7' in 270 means seven tens and not seven hundreds or seven units, because of its position in the number. We also know that there are no units in the number because of the zero written in the units position.

1	I
2	II
3	III
4	IV
5	V
6	VI
7	VII
8	VIII
9	IX
10	X
50	L
100	C
500	D
1000	M

Roman numerals

A chinese abacus. These are still used in many parts of the world

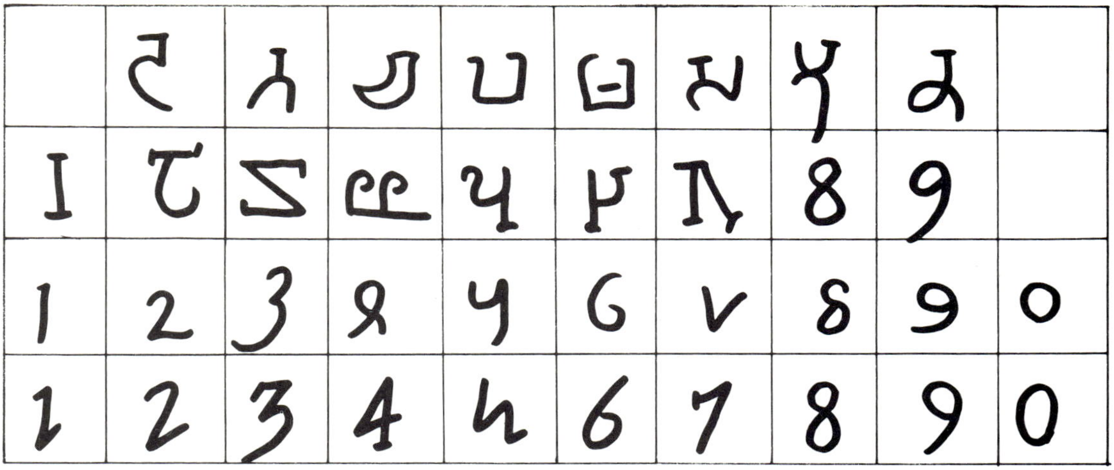

The development of modern numerals from Devangari letters of 2nd century AD

Neither the Romans nor any one else in Europe at that time had worked out the idea of place value or zero. These revolutionary new ideas came from India. A new set of numbers, called the Devangari numerals, were first used about 1200 years ago.

The new numbers were taken up by the world's great trading civilization of the time, the Arabs. Islamic merchants soon saw the advantages of the system, and used it in their trading accounts. Arab scholars used it to develop mathematics, using slightly differently formed symbols. In some ways, these numbers are rather like the numbers we use today.

Also from India, at the same time, came the idea of negative numbers, used to represent debts. The idea of a number line, with numbers moving from negative to positive through zero, was a great development.

The new 'Arabic' numerals meant that pen and paper calculations were possible and many books published in the 14th to 16th centuries showed how to do sums with the new numbers. In England, *The Crafte of Nombrynge*, written in about 1300 AD, was probably based on a translation from Arabic and Hindu sources. It describes how to subtract 134 from 222:

> 'Take four out of two. It will not be,
> therefore boro one of the next figure,
> and set it over the head of the first
> two and reckon it for ten . . . .'

This description has been traced back to the Hindu book *Lilivati*, written by the Indian mathematician Bhaskara nearly 900 years ago.

Arabic numerals

1	١
2	٢
3	٣
4	٤
5	٥
6	٦
7	٧
8	٨
9	٩
0	٠

Hindu mathematicians had also worked out a method of multiplying the new numbers together, using a grid. To multiply 163 by 27 a grid would be drawn with the numbers written at the side.

Each separate pair of numbers was multiplied and written in the box. Two digit numbers were written across the diagonal, for example $6 \times 7 = 42$.

When the whole box was completed the numbers were added up along the diagonals.

This method was sometimes called the *gelosia method. Gelosia* is the Italian word for a grating, and it was called this because the board looked like a grating.

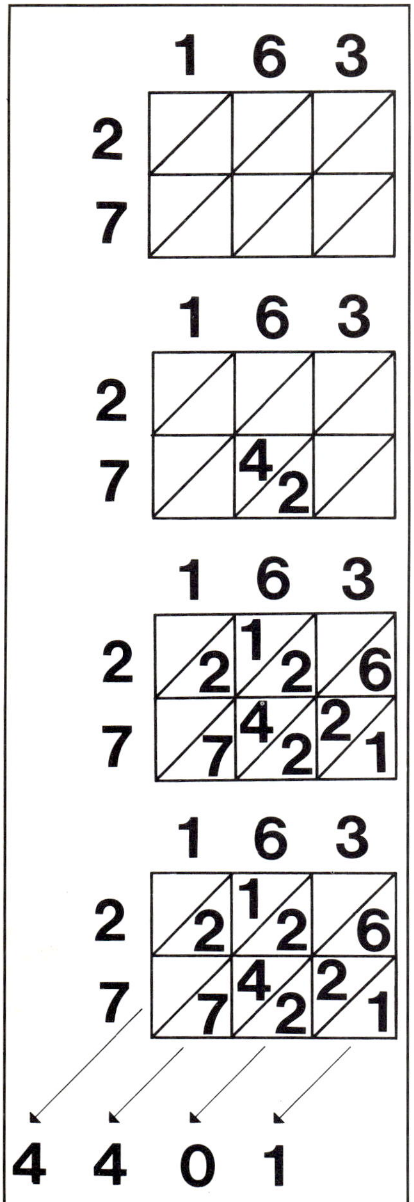

Using the gelosia method

Add the numbers in the diagonals starting from the right. Carry over the tens to the next diagonal

Using Napier's bones

A cylindrical version of Napier's bones

In 1617, the gelosia method of multiplication was simplified by the Scottish mathematician, John Napier. He made a set of nine rods with a table of numbers engraved on each. To multiply 163 × 27, you took the rods numbered 1, 6 and 3 and laid them together with a rod which was marked with the numbers 1 to 9. The answer was then calculated by the gelosia method.

As these rods were often made of bone they became known as Napier's bones.

Italian mathematicians invented many new ways of multiplying numbers together, and by the 16th century there were several dozen methods in use.

Division was more difficult. Only skilled mathematicians ever attempted this. The galley or 'scratch' method involved crossing out each number as it was used. This method was still common in England until about 250 years ago.

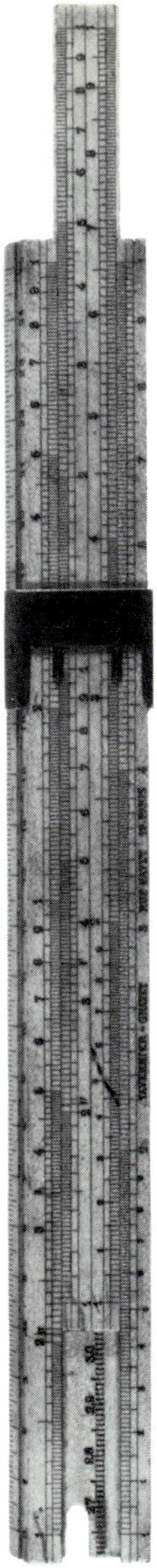

Slide rule

Another great innovation in written mathematics was the idea of decimals. Decimal places were first used instead of fractions at the end of the 15th century. Multiplying and dividing decimals was much easier than using fractions. It was first described by Simon Stevin, in 1585. His decimals did not look quite like ours. Instead of 17.635, he would have written 17(0)6(1)3(2)5(3). Other ways of writing decimals were 17 635 or 17/635.

Using decimals, John Napier was able to extend his calculating tables to deal with very large numbers. These tables were long lists of numbers, called logarithms. Large numbers could be multiplied or divided simply by adding or subtracting a pair of logarithms.

An English mathematician called Gunter invented a device called the slide rule which was based on logarithms. The slide rule was basically a ruler with a sliding middle section from which readings could be taken by matching the numbers on the two scales. Finding logarithms using a slide rule was much quicker than looking them up in tables.

Logarithms and slide rules were not absolutely accurate, but the answers they gave were good enough for most practical purposes and allowed calculations to be made more quickly.

Comptometer

The inside workings of Pascal's calculator

The idea of a calculating machine can be traced back to the French mathematician and philosopher, Blaise Pascal, in the 1650s. His set of cogged wheels and numbers were turned by hand and, though they worked, the idea was not widely used.

In the 19th century, the English mathematician and inventor, Charles Babbage, took up Pascal's idea. He built his first Difference Engine in 1822. This was a set of carefully made wheels that could be used to add, subtract, divide and multiply. The trial version worked, but when Babbage tried to make a full scale machine, he found that the parts could not be engineered finely enough. He abandoned the idea, and worked instead on an Analytic Engine driven by steam. This machine used punched cards to feed in the numbers and information – an idea invented by Jacquard to control patterns in woven carpets. Babbage's Analytic Engine would have been the first computer, but unfortunately the engineering let him down again.

The comptometer was a simpler 20th century invention. It was similar in many ways to the Difference Engine, but was powered by electricity. Revolving notched wheels whirled round to calculate sums. Comptometers couldn't really remember or store numbers, however, and all numbers had to be added by pressing rows of keys. The comptometer was in common use until the 1960s.

Burroughs machine

Comptometers and their rivals, Burroughs machines, were mechanical. Computers are electronic. They work by storing or passing on electrical impulses. Computers can perform complicated calculations very quickly. They do this by breaking problems down into an enormous number of simple steps. Each step is done at great speed.

Computers work in a binary number system, so that the answer to each part of a problem can only be 0 or 1. These numbers act as gates which will, for example, only open if the answer is 0.

The first modern computers were built in the 1940s. In 1943, the Americans built an enormous general purpose computer using electromagnetic relays. It was very slow and made a noise 'like a roomful of old ladies knitting away with steel needles'. Meanwhile, in Britain, the first electronic computers were built to crack German military codes. These machines could do nothing but crack codes.

The first general purpose electronic computer was called ENIAC (the Electronic Numerical Integrator and Calculator). The ENIAC could be rewired to do different kinds of computing.

Early computers used thousands of valves. These were much quicker than magnetic relays, but were expensive and fragile. In the 1950s, the valve was replaced by the transistor which meant that computers could be made much smaller. Even then, a small computer could be the size of several large wardrobes, and a powerful computer might be as big as an aircraft hanger.

In the late 1960s, the microchip was developed. This allowed transistor circuits to be made far smaller. A million parts could be fitted on to a centimetre square wafer of silicon. Using valves, the same million parts would fill a football pitch. It was possible to make cheap, small, simple electronic calculators and full scale computers with large programmable memories.

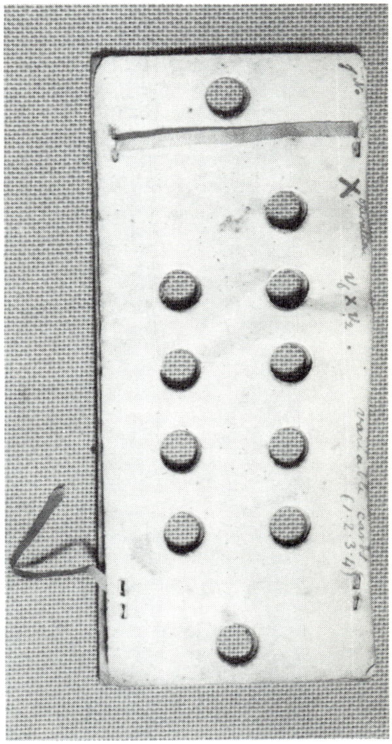

Part of Babbage's Difference engine and a punch card

7 Mathematics today

In a busy society like ours, we are using mathematics all the time. We use arithmetic for shopping, paying bills and checking change. We measure things like ingredients for cooking and use geometry to work out how things will fit together. But more complicated mathematics is going on all around us, helping people to make decisions.

All kinds of information can be recorded in figures and graphs, so that the information can be stored easily and compared. These statistics help people to interpret information and make forecasts. For example, town planners have to work out where new roads are needed. They have to measure how much traffic there is and try to predict how the traffic will change. Will more people own cars in a few years' time, or will more people use public transport? Will more people travel between A and B, or between A and C? Should heavy traffic be diverted from the centre or residential areas of the town? If new housing or new factories are planned, how are they likely to affect traffic patterns?

Statistics are used to sort journeys into different types; geometry helps to solve patterns of distribution and flow of traffic. The plans of the City of Leeds on this page show some of the mathematical possibilities worked out by a planning committee in 1963.

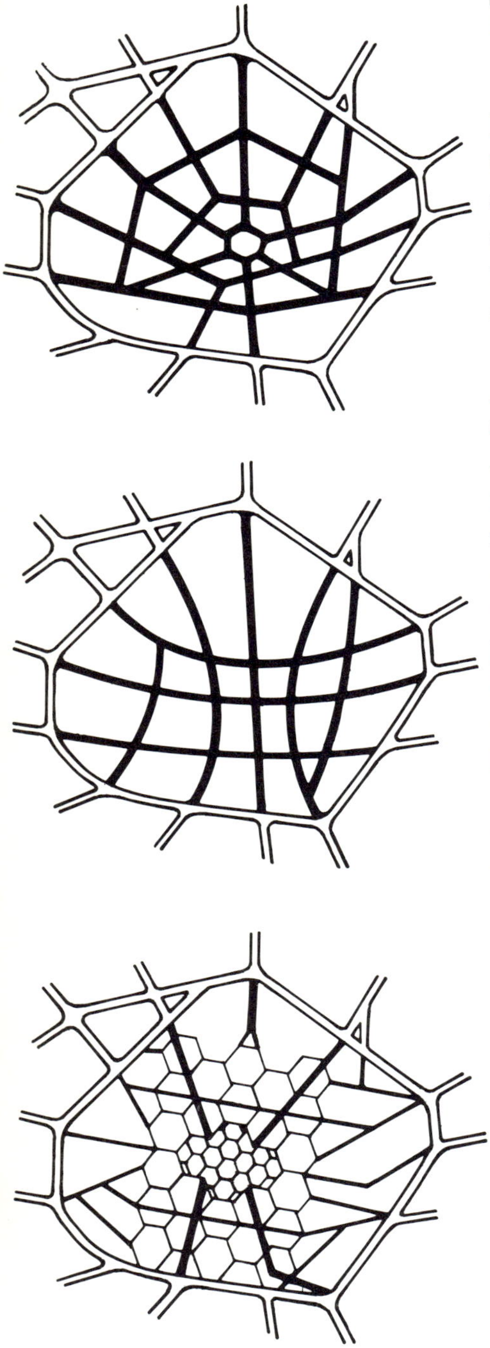

Three different ways of adding to the existing road system in Leeds

Traffic congestion can be avoided by planning alternative routes for road users

Half a million underground passengers were asked about their journeys in this survey

Town planners might also take traffic accidents into account and try to design a road system which will allow fewer accidents. They will try to find out things like what kind of street lighting or sign-posting helps reduce crashes. In Berlin, 2 000 different items of information were recorded for each of 2 000 accidents – a total of 4 million items of information. They included such things as the amount of rainfall, the type of shoes the driver was wearing, and when the vehicle was last checked. A computer was used to store and sort through all this information and quickly answer such questions as 'does the weight of a vehicle in an accident have any effect on the kind of injuries sustained?'.

Road design also calls for a great deal of mathematics to calculate, for example, the likely stresses and strains on a motorway flyover. Engineers can use computers to work out how strong roads and bridges need to be and what materials it would be most economic to use.

Another area where statistics are needed is in planning how schools, hospitals and other services will be used. To do this, the government needs to know where people are living and how old they are. For example, there is no need to build new schools in an area if most of the local people are old. But if there are lots of young couples in an area, it is likely that in a few years time they will want schools for their children and more teachers will be needed, too.

Every ten years the government holds a census. On a certain day, everyone in the country is counted. Every household has to fill in a long form giving details about the people living there and their accomodation. Even people in hotels or away from home have to fill in the form. Then the information is collected from each house by a census returning officer, who passes the information on to the civil service to be analysed.

Planners can then make calculations to find out how society is changing. They can compare the new information with past information to see whether families are getting larger or smaller, whether people are moving away from some areas into other areas and whether the standard of living is rising or falling.

All the information from the census about individuals is kept secret for a hundred years, but the information on living patterns comes out of the computer in the months and years after the census. This helps local and central government to provide the right services for the area they are responsible for.

In Florida, computers are installed in police cars so that information can be obtained instantly on stolen cars and criminal records

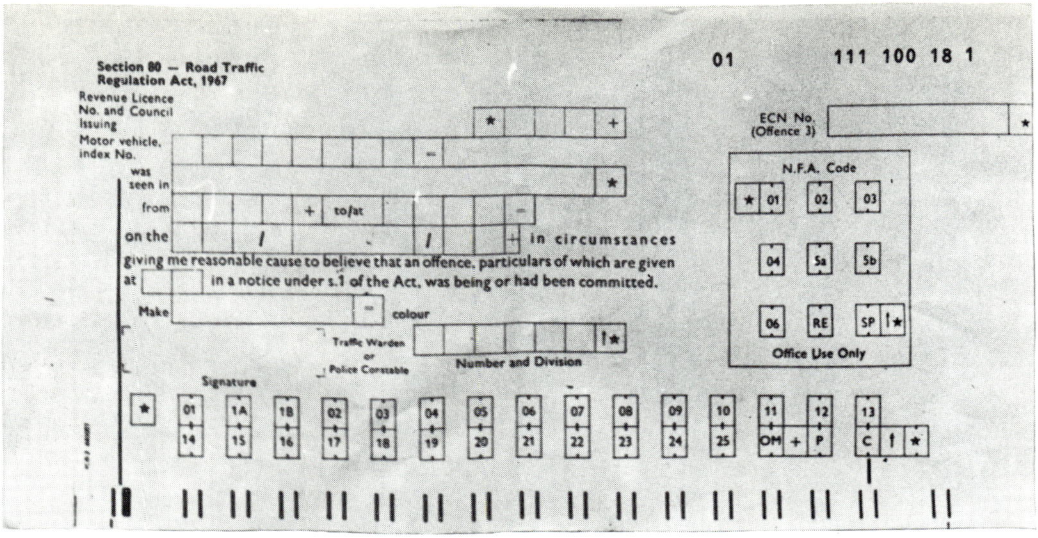

Section 80 — Road Traffic Regulation Act, 1967

01 111 100 18 1

Revenue Licence No. and Council Issuing

Motor vehicle, index No.

ECN No. (Offence 3)

was seen in

from + to/at

on the / / in circumstances giving me reasonable cause to believe that an offence, particulars of which are given at in a notice under s.1 of the Act, was being or had been committed.

Make colour

Traffic Warden or Police Constable

Number and Division

Signature

N.F.A. Code

★ 01 02 03

04 5a 5b

06 RE SP ★

Office Use Only

★ 01 14 1A 15 1B 16 02 17 03 18 04 19 05 20 06 21 07 22 08 23 09 24 10 25 11 OM + 12 P 13 C ★

Parking tickets have numbers which can be read by computer

Today, the medical world is using mathematical skills all the time. For example, a patient with a heart condition is often examined with the help of an electrocardiogram. This machine measures the tiny electrical signals that cross the muscles of the heart. The wave pattern on the display screen shows the doctor what condition the heart is in.

Mathematical techniques are also used to test new medicines. After a new drug has been tested for safety, it is sometimes tried out on patients to see if it actually makes them better. Some patients get better just because they know they are taking part in an experiment. Other patients improve because they are taking something that looks like medicine. So it is difficult for medical researchers to know if a new drug works or not.

To overcome this problem, doctors give half the group of patients the new drug. The other half of the group are given a similar-looking but harmless 'drug' called a *placebo*, that won't cure the patient.

Both groups of patients must have the same proportion of young and old, male and female, and so on, because it would be unfair to test the new drug on a group of young women and the placebo on a group of men in their eighties. Statistical methods are used to select the patients in each group.

The doctors giving the drugs and examining the patients for improvements are not told by the researchers which is the drug and which is the placebo. Using mathematics, the results of the experiments are carefully checked to see if the difference between the two groups could be due to chance.

Launch control at Kennedy Space Centre

Many of these mathematical techniques are based on the maths we learn in school. The computer has taken over many of the boring parts of mathematics in lots of jobs. Computers can work faster and more accurately than we can, and calculations that would once have been too complicated or not worthwhile can now be done by the computer.

Rather than risk lives and expensive aircraft, computers are often used to help train pilots. A flight simulator can give the pilot made-up information about things like fuel levels, aircraft weight and air pressure. The trainee pilots use this information to make decisions as though they were on a real flight. During the flight simulation, the computer can adjust the information. For example, as the fuel is used up, the weight of the aircraft will go down. The computer makes these calculations so quickly that the flight simulation is very realistic.

In space flight control, computers are used to monitor all the equipment and make decisions about a lot of the routine work of the spacecraft. This is because there is too much information for the human flight controllers or astronauts to deal with. They need to keep their minds free for any emergency that might arise.

Computers are used every day in shops and supermarkets. Often, the sales assistant will not just ring up your bill at the till, but will also press code keys, or use a light pen on a bar code on the label. This lets the shop's computer know which item you have bought. The computer can then calculate quickly how many items are left in the shop, and order more if they are needed. In the past, a stock controller would have had to count how much stock was left on the shelves at the end of each day and, at the end of each week, order replacement stock. With the computer, the replacement goods may already be on their way to the shelves before you've left the shop!

Modern weapons are often controlled by computer

8 Playing with maths

Mathematics has developed a great deal since we first needed to count. It is something we use every day and often take for granted. But the theories behind some of the mathematics we use can be very complicated. Here are some teasing problems that you might have fun trying out.

Soma cubes

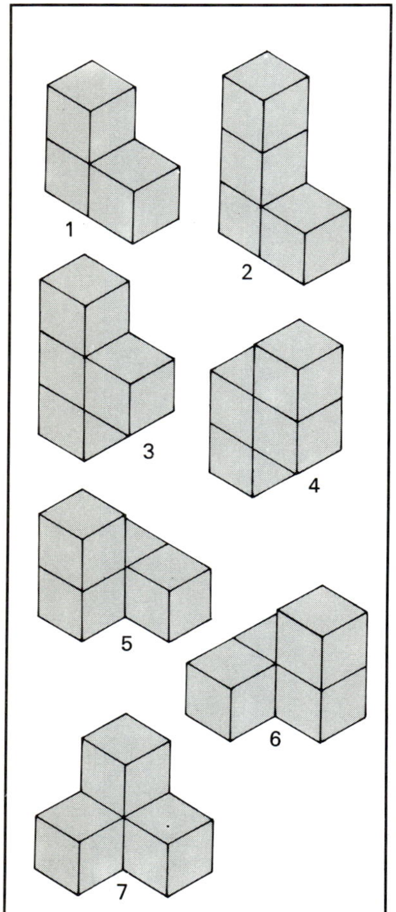

1 irregular shape using three cubes
2–7 irregular shapes using four cubes

All the shapes on the right can be made from these seven soma cubes

Soma cubes were invented about twenty years ago by the Danish mathematician and writer, Piet Hein. Using no more than four cubes, he made up all the irregular shapes possible. By 'irregular' he meant that each shape should have at least one corner facing inwards. Only one irregular shape could be made with three cubes. There are six possible shapes with four cubes. Two of these shapes are mirror images of each other. The interesting part is that all these shapes fit together to make up a larger cube. But there are many more shapes which can be made with these seven Soma cubes.

The Mobius band

Take a long strip of paper. How many sides does it have? How many edges does it have?

Stick one end of the strip to the other to make a loop. Now count the sides and edges again.

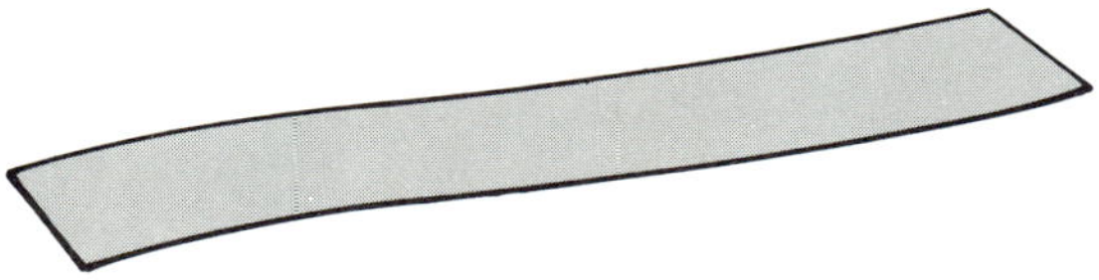

1

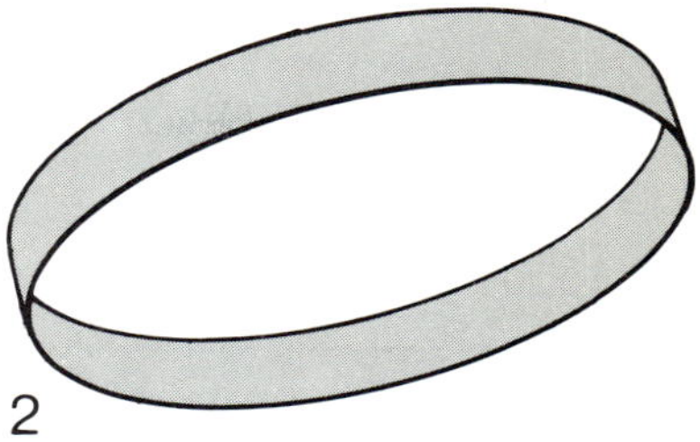

2

Can you imagine a strip with only one side and only one edge? Take a second strip, and this time give it a single twist before you stick the ends together.

This is a Mobius band. Try drawing a line along one side of the strip of paper. Now cut the Mobius band in half along its length. What do you think you will make? Is it still a Mobius band?

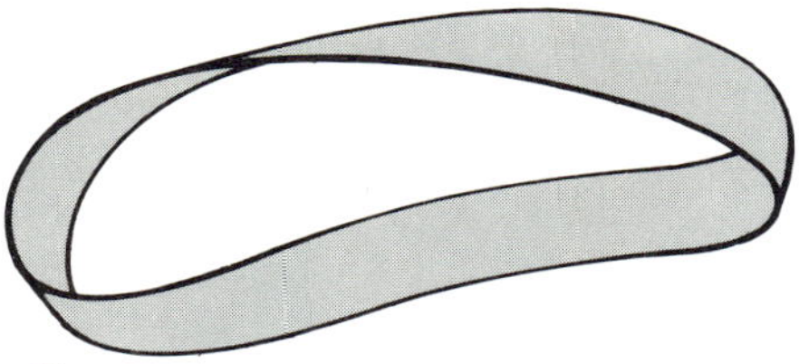

3

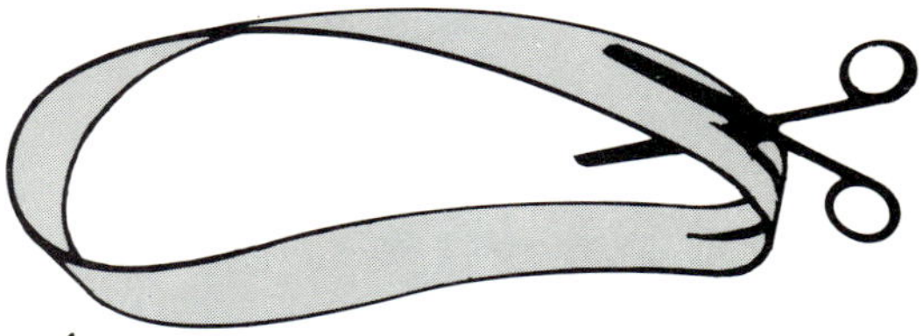

4

Make another Mobius band, and cut it one-third of the way across the width all the way around – and you will get another unexpected result.

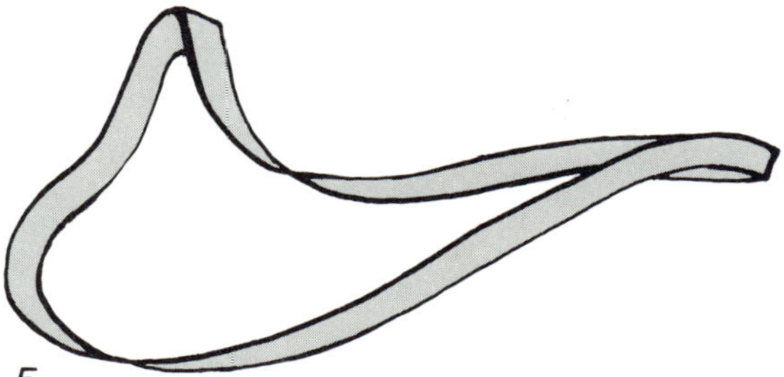

5

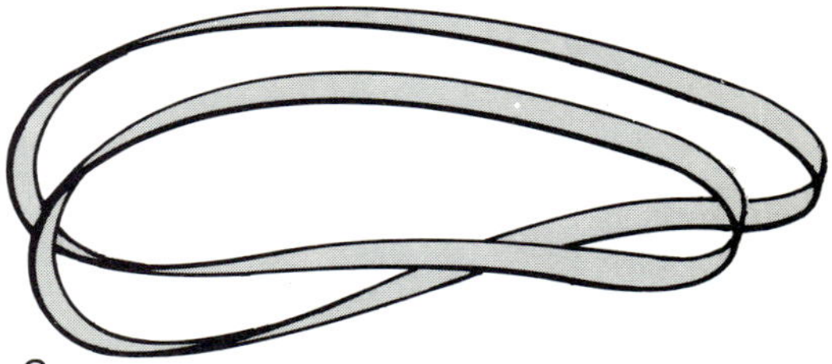

6

Try making a double Mobius band, with two strips of paper. Do you have two Mobius bands? Run your finger between the two to find out. Then try separating the two.

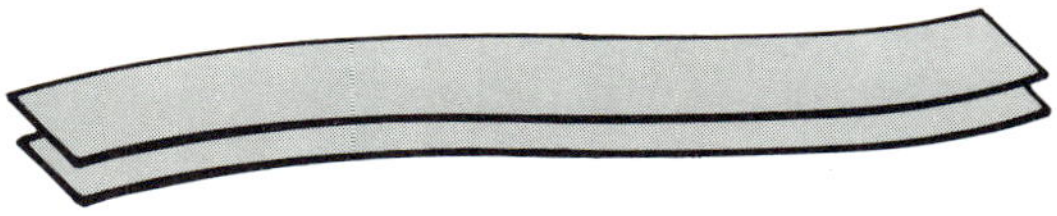

7

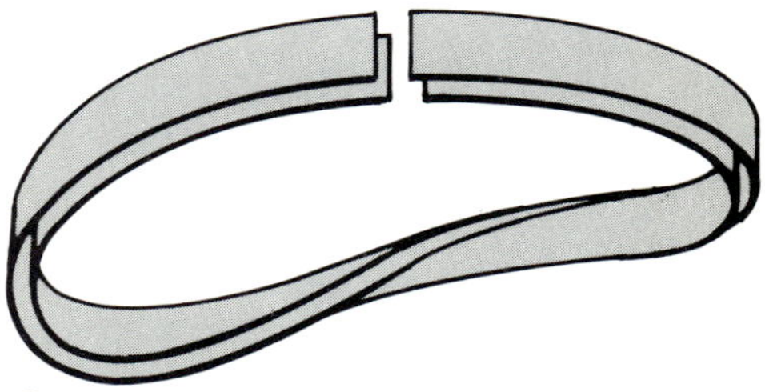

8

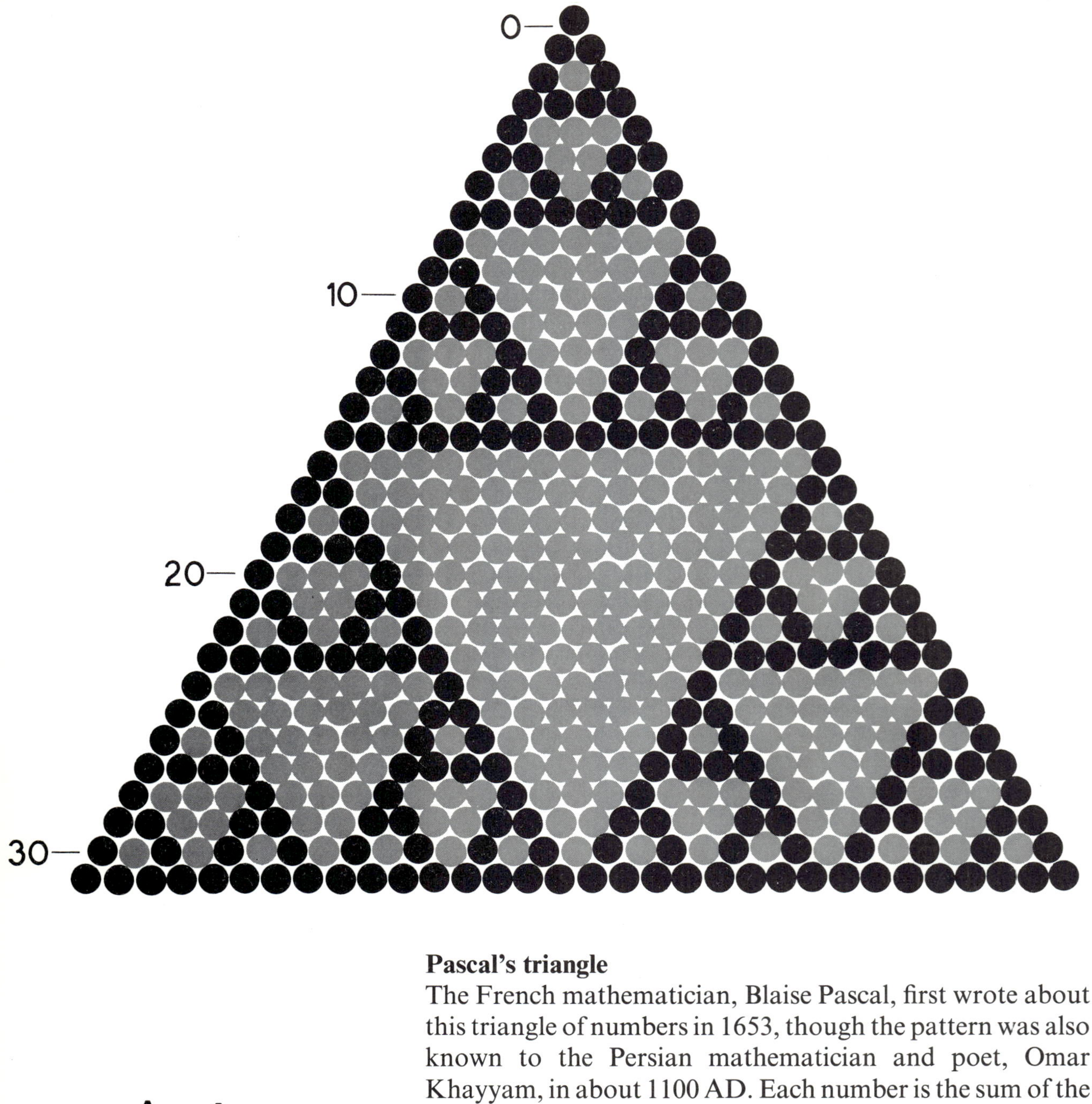

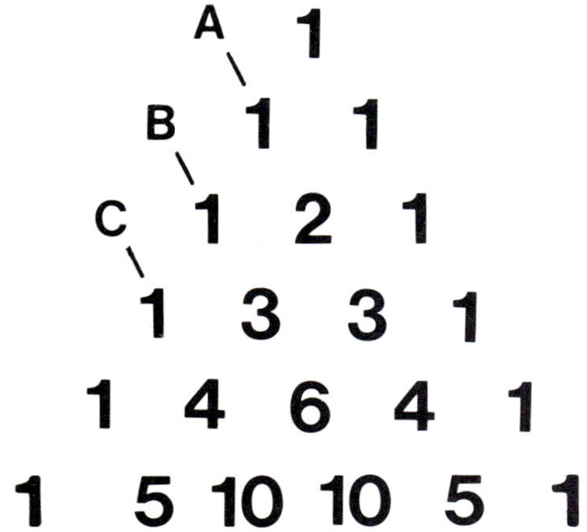

Pascal's triangle

The French mathematician, Blaise Pascal, first wrote about this triangle of numbers in 1653, though the pattern was also known to the Persian mathematician and poet, Omar Khayyam, in about 1100 AD. Each number is the sum of the two numbers immediately above it.

The triangle can go on for ever. There are an amazing number of patterns to be found in the triangle. Diagonal A, for example has all the numbers in order. Diagonal B gives every triangle number, that is the number of points that can be arranged in rows to form a triangle. Adding pairs of triangle numbers gives square numbers ($3+6=9$, $6+10=16$, etc). Diagonal C gives all numbers that could make up the sides of different shaped pyramids. Colouring in odd and even numbers gives the pattern above. All sorts of patterns can be found in this triangle of numbers.

The towers of Hanoi

This game, invented in 1883 by the French mathematician, Lucas, involves eight different sized discs stacked on a peg. The object of the game is to move the stack from one peg to another. However, only one disc may be moved at a time, all the discs must be on one of the three pegs, and no disc can be put on top of a smaller disc.

How many moves are needed? If you added just one more disc to the stack, how many more moves would be needed?

The original game was supposed to be based on a Tower of Brahma in Varanasi, India. There, a tower made of 64 discs of gold was being moved by priests, and it was said that when they had completed their task, the world would end. The story isn't true, but it would take 18,44,744,073,709,551,615 moves to rebuild the tower!

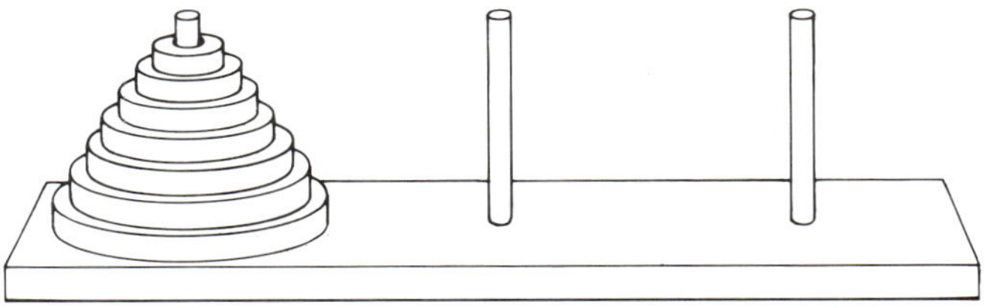

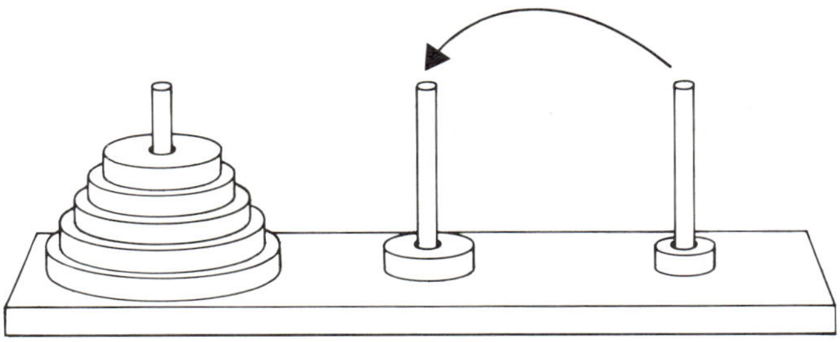

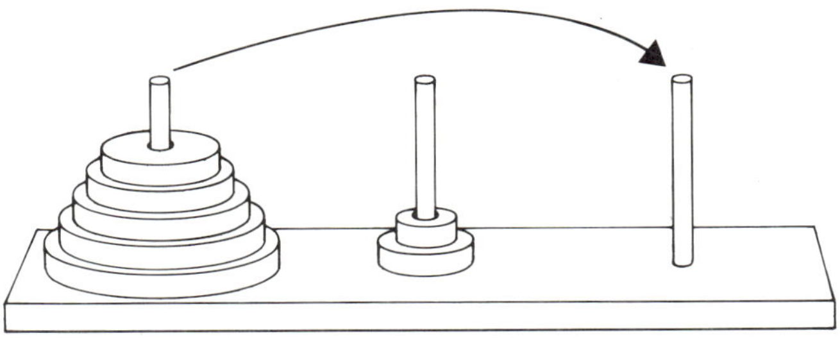

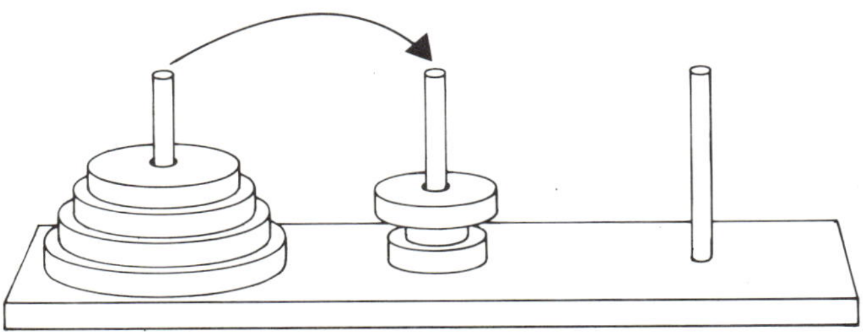

This move is not allowed

This magic square is in Dürer's *Melancolia*

Magic Squares

The oldest known magic square originally came from China. The 3 000 year old book, *Yih King*, shows a pattern which is said to have come from the markings on a turtle's back. In the magic square, each number, from one upwards, is written down in a grid, so that each side and each diagonal add up to the same number. The pattern is still thought to be a charm in the far east.

No one has yet found another magic square for a 3 × 3 grid. But there are many possibilities for a 4 × 4 grid. Each side should total 34.

This example was drawn by Albrecht Dürer, and has three other interesting points: It is made up from the consecutive numbers 1 to 16; adding up the two diagonally opposite numbers always gives 17; and the two numbers in the centre of the bottom row give the date that Dürer made the picture, 1514.

The oldest known 4 × 4 magic square is from an 11th century carving at Khajuraho in India.

This is even more curious. The order of any row or column can be changed without altering the totals.

This 8 × 8 magic square was made by the 18th century mathematician Leonard Euler. It can be split into four separate magic squares, and if you were to move a chess knight across the grid, you would move from square one to 2 to 3 up to 64 without visiting a square more than once.

3 × 3 cube from the *Yih King*

4	9	2
3	5	7
8	1	6

4 × 4 cube from Khajuraho

7	12	1	14
2	13	8	11
16	3	10	5
9	6	15	4

Euler's square

1	48	31	50	33	16	63	18
30	51	46	3	62	19	14	35
47	2	49	32	15	34	17	64
52	29	4	45	20	61	36	13
5	44	25	56	9	40	21	60
28	53	8	41	24	57	12	37
43	6	55	26	39	10	59	22
54	27	42	7	58	23	38	11

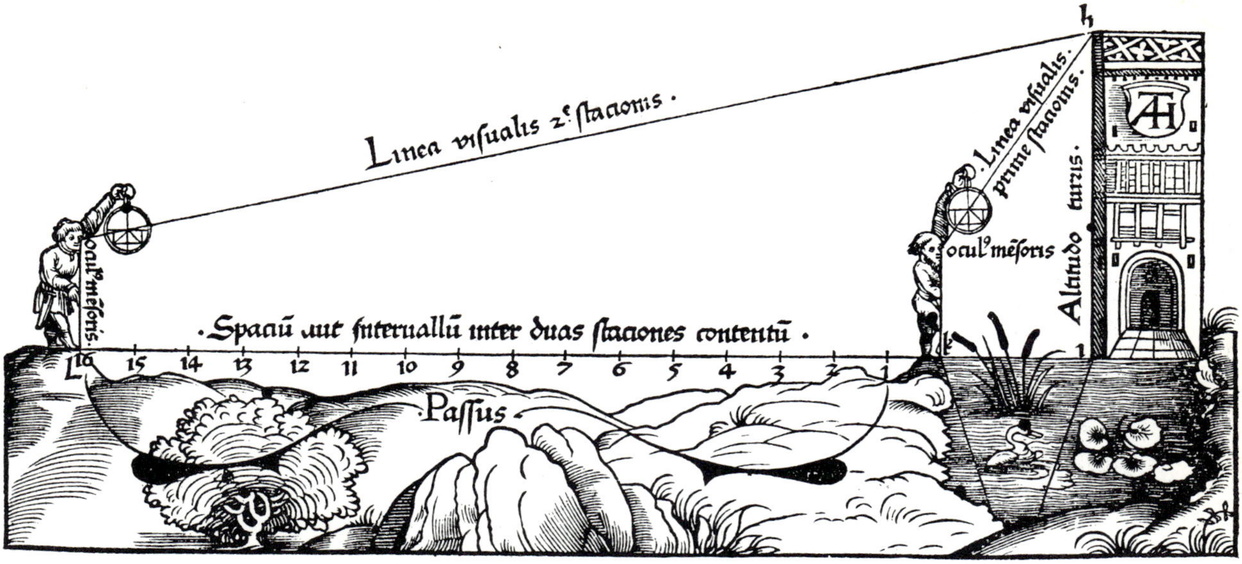

Using an astrolabe to measure the height of a tower, 1513

More books to read

Count on your Fingers African Style by Claudia Zaslavsky (Harper & Row, 1979)
Tic Tac Toe, and Other Three in a Row Games by Claudia Zaslavsky (Harper & Row, 1982)
Think of a Number by Johnny Ball (BBC Publications, 1979)
Famous Names in Science by Paul Clark (Wayland Publishers, 1979)
Mathematical Carnival by Martin Gardiner (Penguin Books, 1978)
Mathematics for the Million by Lancelot Hogben (Pan Books, 1967)
Mathematical Games and Activities by Roy Hollands (Hart-Davis Educational, 1977)
Number Puzzles edited by Kelsey (Usborne, 1980)

Acknowledgements

Aerofilms, page 11; Barnaby's Picture Library, page 52; British Library, details page 34, page 46; British Museum page 5, 8; Courtauld Institute, title page, page 25; Mary Evans Picture Library, pages 12, 26, 27, 31, 32, 39, 43; Werner Forman Archive, pages 7, 8, 9, 33; John Freeman & Company, pages 15, 24, 28; Richard and Sally Greenhill, front cover (top), page 4; Ronald Sheridan, front cover (bottom), biblio page, pages 6, 13, 16, 31, detail page 35, 35, 39; Keystone Press Agency, pages 53, 54, 55, 56, 57; John Mansell Collection, details pages 14, 15, pages 20, 21, 22, 23, detail pages 23, 62; Ann Ronan Picture Library, pages 38, 42; British Crown Copyright, Science Museum, contents, pages 10, 17, details pages 17, 18, pages 27, 28, 29, 30, details pages 36, 37, pages 38, 44, 47, 48, 49, 50, 51, 63.

Artwork by B L Kearley Ltd.

We are grateful to the Controller of Her Majesty's Stationery Office for permission to use three diagrams from Buchanan Report, page 52.

Index